What Is Life?

Also by Ed Regis

The Info Mesa

The Biology of Doom

Virus Ground Zero

Nano

*Great Mambo Chicken and the
Transhuman Condition*

Who Got Einstein's Office?

What Is Life?

Investigating the Nature of Life

in the Age of Synthetic Biology

ED REGIS

OXFORD
UNIVERSITY PRESS

OXFORD
UNIVERSITY PRESS

Oxford University Press, Inc., publishes works that further
Oxford University's objective of excellence
in research, scholarship, and education.

Oxford   New York
Auckland   Cape Town   Dar es Salaam   Hong Kong   Karachi
Kuala Lumpur   Madrid   Melbourne   Mexico City   Nairobi
New Delhi   Shanghai   Taipei   Toronto

With offices in
Argentina   Austria   Brazil   Chile   Czech Republic   France   Greece
Guatemala   Hungary   Italy   Japan   Poland   Portugal   Singapore
South Korea   Switzerland   Thailand   Turkey   Ukraine   Vietnam

Published by Arrangement with Farrar, Straus and Giroux, LLC

First published as an Oxford University Press paperback, 2009
198 Madison Avenue, New York, NY 10016

www.oup.com

Oxford is a registered trademark of Oxford University Press

Library of Congress Cataloging-in-Publication Data
Regis, Edward, 1944–
What is life? : investigating the nature of life in the age of
synthetic biology / Ed Regis.
p.   cm.
Originally published: New York : Farrar, Straus and Giroux, c2008.
Includes bibliographical references and index.
ISBN 978-0-19-538341-6 (pbk.)
1. Life (Biology)—Philosophy.   2. Vital force.   I. Title.
QH501.R434   2009   570.1—dc22   2009003487

1 3 5 7 9 8 6 4 2

Printed in United States of America
on acid free paper

*The author gratefully acknowledges a research grant from
the Alfred P. Sloan Foundation.*

For Pamela Regis

Contents

What Is Life?

The Second Creation

IN THE SUMMER OF 2002, in northern Italy, a group of three scientists and a philosopher agreed to create a new form of life. The four were old friends, buddies, and during the course of an annual reunion they realized that at this precise stage of their respective careers each of them wanted to be working on some important, ambitious, and challenging new project.

So they decided they would create life. Not a simulation of life. Not an imitation of life. Not fake life. Rather, real life, a genuinely new living entity, albeit one not based on biology and not made out of the customary biological ingredients: no DNA, no conventional biomolecules, no cell membrane of the ordinary type, no nucleus, no mitochondria, no endoplasmic reticulum or any of the other innumerable vital trappings of normal, orthodox biological cells.

None of the group members was sure what the exact definition of life was, if indeed there was one, although metabolism, self-reproduction, and the ability to evolve seemed to be essential defining criteria of any entity that was to be classified as alive. For the time being, however, the obstacles to creating new life were not definitional, but rather scientific and technical.

For one thing, creating life from scratch had never been done before—at least not by scientists. Life on earth had a genesis, that was for sure, but what the exact order of events had been that led to the origin of life was, to put it mildly, an unsettled question in biology.

The prospect of creating new life also raised some entirely novel technical questions. What kind of food would an artificial organism eat? Would new life come into existence all at once, like a bolt of lightning, or gradually, in stages? Would it need some sort of chemical weaning process, a form of artificial life support?

Where would the money come from? What lab or labs would be involved, and in what countries? Could a passably blasphemous project such as this be attempted in America, many of whose citizens were unduly alarmed even by the idea of genetically modified foods? What were the potential risks involved? Was a population of synthetic living entities going to be dangerous, a threat to civilization?

The project would be no trivial undertaking, but that only added to its appeal: there was no glory in doing something easy. Creating life would be, as Robert Oppenheimer had said of building the atomic bomb, "technically sweet," a goal scientifically so tempting as to be almost irresistible.

The group had no guarantee of success, but they had several things going for them, the most important of which was that life had already emerged once: in other words, they had a proof of concept. Second, modern science had an extremely good understanding of how life worked, right down to the smallest molecular details. And so conventional life, its general architecture, mechanisms, and arrays of vital processes, could be used as models for new life-forms, examples of which the researchers would produce in different media and with different raw materials.

Third, science had ascertained with the utmost precision how certain chemicals reacted with others and how a given type of molecule could hold information and control the activities of other molecules, or groups of them—all of which meant that the current level of chemical knowledge was potentially up to the task of building a living, metabolizing synthetic cell. Fourth, machine technology had reached the point where miracles of manipulation could be performed by devices that worked at the very finest physical scales, which was a capacity the scientists could exploit in the creation of their microscopic artificial organisms. Fifth and last, there was the computer. By this point, practically any envisioned process, entity, or anything else could be simulated beforehand, in arbitrary detail, and repeated as many times as you wanted—previewed, reviewed, revised, rewound, run backward—all to serve as a guide to the physical realization of the phenomenon in question. And so it should be possible to simulate, well ahead of building it, the full developmental path and the complete inner workings of an artificial living cell.

In the progress of science and technology there has often been a distinct, ripe psychological moment at which a given

6 · WHAT IS LIFE?

revolutionary undertaking that was previously impossible becomes suddenly doable—a project uniquely suited to the
times. And at this precise moment in history, the time was
right for creating life. The scientists might fail in their attempt, but surely others would succeed, for there was unlikely to be any insuperable barrier to the ultimate realization
of their goal.

In any case, success, or even failure, would raise an additional bunch of questions, ones that reached far beyond the
narrowly technical issues. What were the stakes here, scientific, moral, political, legal? The problems started with the
perennial and trite layman's taunt, the claim that creating life
was "playing God." That comment had been spat out as an
insult aimed at nearly every radical scientific advance, from
splitting the atom to birth control to genetic engineering, organ transplants, and whatever else. Still, the question would
have to be faced.

Behind the layman's taunt, however, was a cluster of legit-
imate concerns. Would the attempt to create new life venture
into forbidden territory, sacred, untouchable realms into
which mere mortals shouldn't even think of entering? Had
science, in other words, finally reached its proper stopping
point?

What light would a new form of living matter shed, if any,
on the "meaning" of life—whatever was to be understood by
that mystifying phrase? What would it say about the worth
and uniqueness of human life, or of life on earth, human as
well as animal? Would human life, or any form of life, have
less value if we could create new life-forms at will, like works
of art? Were animal rights activists going to claim that these

new life-forms had rights, too? Would new life be more like medicine . . . or more like poison?

For that matter, a collection of hot-button, lightning-rod issues swarmed around the concept of life like bees around a hive: the problem of abortion, for example, which rested in part on the question of when human life began. At the other limit of human existence, the issue of euthanasia, the use of heroic measures to artificially prolong life, and when to withdraw them, were bound up with the determination of when life, in any meaningful sense, ended. There were the further problems of our moral obligations to other species, especially to those endangered species, such as insects, that might be considered "lower" forms of life; whether the harvesting of embryonic stem cells constituted the taking of a human life; whether an advanced form of artificial intelligence, if and when it ever existed, would itself be a new kind of life-form; and so on.

The scientists involved in creating new life were not motivated by a desire to answer such questions—nor had they any special competence for doing so. Science, after all, was concerned with what is, not with what ought or ought not to be. The most profound and provoking question raised by the effort to build an artificial living cell, therefore, was exactly the one that lurked as an unseen presence beneath all the rest: the age-old riddle, What is life?

One

Birth of a Cell

MAY 2005. In a new industrial park at Porto Marghera, some four miles across the lagoon from Venice, an American physicist by the name of Norman Packard is staring at the enormous 30-inch-wide display screen of a Macintosh G5 computer. Floating around against a dark background is a dense assortment of red, green, and blue dots.

"Blue is water, the greens are hydrophobic molecules, which means they don't like water, and the reds are hydrophilic molecules, which do," Packard says.

The simulation begins with the dots spread out evenly across the screen in a relatively homogeneous mix. But then in the incremental time-steps of the particle dynamics program, a pattern emerges. The greens move toward one another and then converge and clump together, forming a spherical structure. The reds, meanwhile, follow the greens and arrange themselves on the outside of the mass, as if to protect it from intrusion. The result is a vesicle, a tiny bilayered fluid-filled sac. The vesicle has formed itself spontaneously, the result of a self-assembly process driven by Brownian motion (the random thermal movement of molecules in a fluid medium) and by various chemical reactions.

"We believe that this combination of chemical reactions and self-assembly is one of the crucial combinations that we need to understand to make these artificial cells," Packard says.

Artificial cells? Venice? A city of more than a hundred churches, miles of canals, and innumerable ancient palazzi, all of them suspended in time, a place where nothing fundamentally new has happened for hundreds of years? Somehow the location is strangely fitting. In its heyday, Venice was a world-class power and trading center as well as a realm of considerable intellectual freedom. The city was now and always had been home to a variety of creative spirits: composers, artists, and scientists, including Galileo. And its labyrinthine streets and alleys were bathed in the green waters of the Venetian lagoon—water just coincidentally being the medium in which, according to most theories, earthly life originally began. So why should it not begin again, here?

Norman Packard, for one, finds no incongruity in the prospect. Packard is the chairman, CEO, and scientific head of ProtoLife s.r.l., a Venetian start-up company located in Parco Vega, a technology park the regional government had created on the grounds of an old chemical factory.

"The city of Venice, but even more generally the region of Veneto, wants to diversify its portfolio of activities," Packard said. "Venice has this very strong component of tourism that dominates its economy in many ways, and so it's trying to create some economic diversity that can give a certain kind of life to the city, not related to tourism."

ProtoLife's business plan is founded on an attempt to start life over, to begin from the beginning. It's not their intention to redo Genesis, outdo Frankenstein, or to blaze a path of glory through one of the final frontiers of applied science— although, if they're successful, Packard and his crew will end up doing all those things. The company's motivation is far more prosaic, practical, and commercial: to create artificial cells. Made from scratch and called "protocells," they will be programmed to carry out useful tasks such as synthesizing vaccines and drugs, cleaning up toxic waste, scavenging excess CO_2 from the atmosphere, and other such miracles, and earning the company a tidy profit in the process.

After watching his simulation run a few more times— "We've done between six and seven thousand runs so far," he says—Packard walks down a polished green marble hallway, turns right, unlocks a door, and enters the company's lab suite. This is the domain of ProtoLife's chief chemist, Martin Hanczyc, a postdoc Packard recently hired away from Jack Szostak's competing artificial cell project at Harvard. In fact, ProtoLife is only one of a half dozen or so scientific efforts bent on creating new life: in addition to the ProtoLife and Harvard projects, there are others at Rockefeller University in New York, the University of Nottingham in England, and the University of Osaka in Japan, among other places. All too obviously, creating life is an undertaking whose time has come.

Hanczyc's laboratory at ProtoLife boasts a full supply of chemical apparatus: the usual lab glassware, serological pipettes, fume hoods, scales, centrifuges, microscopes, plus heavier machinery. "This is one of our main analytic tools, a combination spectrophotometer and fluorometer," Packard says of a large piece of equipment. "You find this in practically every chemistry lab in Europe, so we have one too."

Hanczyc has been synthesizing and studying various types of vesicles, and today Packard wants to show me what they look like. Packard is a big man with shaggy blond hair, glasses, and a courtly manner. He has a slow and deliberate style of speech, which includes a precise, mellifluous Italian, courtesy of his wife, Grazia Peduzzi, who was born in Milan. He squints through a fluorescence microscope, adjusts the focus, and finally, there they are: the real-life correlates of the objects he had been simulating on the computer.

"Somewhat dried up," he says of the vesicles, which Hanczyc had prepared a while ago.

A vesicle is not a living thing. It's just a shell, a husk, the merest framework of the full artificial cell that's supposed to assemble itself on the premises and spring into life at some undefined point in the future. Nevertheless, what we have here on the microscope stage is something passably astonishing, slight and rudimentary though it might appear at first glance. For these filmy minute blobs are the first stirrings of an event that last took place billions of years ago: the genesis of life.

THE DREAM OF creating life has ancient roots in the human imagination. In *Frankenstein*, which Mary Shelley completed in 1817 at the age of nineteen, the scientist Victor Franken-

stein cobbled together a creature from body parts he'd spir-
ited away in the dead of night from graveyards, dissection
rooms, and slaughterhouses. The resulting beast came to life
when Dr. Frankenstein, by unspecified means, infused "a
spark of being into the lifeless thing that lay at my feet."

Serious scientific attempts at infusing a "spark of life" into
inanimate flesh go back at least to Luigi Galvani's discovery
in 1771 that by applying electrical currents to a dissected frog's
legs he could cause them to twitch as if alive. A hundred years
later, in 1871, Darwin spoke of life as possibly having arisen
"in some warm little pond, with all sorts of ammonia and
phosphoric salts, lights, heat, electricity, &c., present."

As if following Darwin's recipe, when twentieth-century
scientists approached the problem of understanding how life
originally arose on earth, they attempted to re-create what
they thought were the original prebiotic conditions. The
canonical effort, now a cliché of twentieth-century science
history, was the 1952 "Urey/Miller experiment," in which the
chemists Harold Urey and Stanley Miller put ammonia, hydrogen, and methane inside a closed flask, circulated steam
through this "atmosphere," and added bolts of "lightning" in
the form of periodic electrical sparks. All they got for their
trouble were some amino acids (building blocks of proteins)
that were not in the mixture to begin with. The Urey/Miller
experiment was once considered a very big deal, but it isn't by
some of the protocell project's scientists: "We are not search-
ing in the black and hoping that something happens," says
the protocell researcher Uwe Tangen. "We're really trying to
engineer these things."

Attempting to build an artificial cell is hardly a new idea in biology, but the specific protocell design Packard and Hanczyc are working on originated with Packard's longtime friend, the Los Alamos physicist Steen Rasmussen. Even as a boy in Denmark, Rasmussen liked to grapple with the big questions. He was by nature of a metaphysical turn of mind, and while still a kid he discussed subjects of cosmic import with his father, who was a bricklayer. Did the universe have a beginning—or an end? Where did it come from? Where was it going?

Later, in the 1980s, Rasmussen, together with Chris Langton, Norman Packard, and some others, became one of the founding fathers of the artificial life (ALife) movement. Launched at a Los Alamos workshop in 1987, artificial life was an attempt first to simulate and then actually to create a new life-form. Supposedly there was to be "soft," "wet," and "hard" artificial life, existing in the form of software, wet chemistry, and robotics, but the reality of the situation turned out to be quite different. "Most of the activities in the artificial life community have been with simulations," Rasmussen admits.

For a long time, that was true even of Rasmussen himself, who over the years had run countless computer simulations of various life-forms, modeling their possible self-assembly routes, evolutionary development pathways, and so on. But his abiding passion had always been to understand what life was and how it arose. At length he decided that the best way to understand life was to make some of it himself, ab initio.

In truth, he became obsessed with the idea. Although he lived in an adobe-style house surrounded by a number of natural life-forms, including his wife, Jenny, and three kids—not to mention horses, chickens, a parakeet, dog, and cat—he did most of his thinking at the Los Alamos lab and on his daily commute to and from, a route that took him past some of the most inhospitable, sun-blasted terrain imaginable: desiccated red cliffs, dry desert sands, and, occasionally, the whited bones of dead animals.

So daunting was the goal of creating new life, he realized, that only the simplest and most radically stripped-down design would have the remotest chance of actually working. Any given entity, in Rasmussen's view, had to have three main attributes in order to be considered alive: it had to take in nutrients and turn them into energy, meaning it had to have a metabolism; it had to reproduce itself; and its descendants had to be able to evolve by means of natural selection. A conventional biological cell, which did all that and more, was a masterpiece of complexity: it had an outer wall through which various essential substances were selectively transported in and out. It had an inner wall around the nucleus, which did the same. And both the nucleus and the cytoplasm surrounding it were brimming with all sorts of enzymes and other biochemicals, plus microstructures and organelles: the ribosomes, the mitochondria, the Golgi bodies, and all the rest. So very complex were even the simplest biological cells that it was a wonder they worked at all.

Rasmussen didn't want to get bogged down with all that incredible complication and detail, so he set about devising "the most lousy, simple, self-replicating, autonomous unit you can imagine," a cell so small it would be "the size of dust."

He got rid of the DNA, the nucleus, the organelles, and much of the rest of standard cell wetware. His protocell would be a *minimal living entity*, thousands of times smaller than a biological cell, and would be composed of three main structures:

a container made of fatty acid molecules; a primitive metabolic system; and a new type of genetic material called PNA.

Fatty acids were also known as lipids, and their major attraction for Rasmussen was that "they make the containers for free. You put them in water and they make the containers. That's the state they want to be in. They want to join up and make these structures." The component molecules did this on account of their chemical polarity: one end of the molecule was hydrophobic (or water-avoiding), the other hydrophilic (or water-seeking), and so when placed in water the molecules naturally arranged themselves into little sponge-like vesicles with the hydrophilic ends forming the outside surfaces and with the hydrophobic ends huddled together on the inside. (Many of the protocell's activities would be governed by the twin forces of hydrophobia and hydrophilia.)

For genes, Rasmussen needed a molecule that could both contain hereditary information in the manner of DNA or RNA, and could replicate, but without having to go through all the biochemical, biomechanical, and other enzyme-driven contortions those molecules underwent in natural cells. What he needed, in short, was a coding molecule that could unzip and replicate in some quick and dirty, no-sweat, E-Z fashion. For this he chose PNA, peptide nucleic acid, a substance synthesized in 1991 by Peter Nielsen, the Danish biochemist. This was a double-stranded molecule that could split down the middle, just like DNA, uncovering its A, T, C, and G bases. Its advantage for Rasmussen, however, was the different ways in which the double-stranded and single-stranded versions of PNA behaved in the cell. A double-stranded stretch of it was hydrophobic and would sink down into the interior of the container and away from the water that surrounded the cell. At a preset temperature, the PNA molecule would spontaneously separate lengthwise inside the cell. The bases of the two single strands were hydrophilic and would therefore rise up to the cell's outer surface. There they would encounter matching PNA fragments that were also floating in the surrounding water, placed there with malice aforethought by the experimentalists. Those fragments would now attach themselves to the single strands, thereby forming new double-stranded molecules which, hydrophobic once again, would sink back down into the cell's innards. That took care of gene replication.

The protocell's metabolic, growth, and self-reproduction processes would be a product of light-sensitive lipid molecules being force-fed into the container. Light would activate the polarity of the molecules in such a way that their hydrophilic ends would rise to the container's surface and squeeze themselves in and among the other molecules that made up the cell's exterior. When the quantity of those surface molecules reached a certain critical mass, the forces holding them together would be overcome and the cell would split in half, reproducing itself.

Natural selection would come into the picture as the protocells reproduced: those that possessed some selective advantage in the speed or efficiency of replication would displace and ultimately wipe out those deficient in those qualities.

That was the basic design plan and operating formula of Steen Rasmussen's protocell. An ingenious design by any standard, especially if it worked. But in order to put his plan into effect, the wee matter of funding had to be addressed.

"I am doing this because I want to understand what life is," Rasmussen said. "That's the driver. Now that's not enough to get money, so the secondary driver is of course, Well, how can this be useful?"

From a practical point of view, there were three key benefits to Rasmussen's protocells. One was their relative safety: because artificial cells would be structurally and chemically alien to modern biology, they would be far less risky to experiment with than genetically engineered biological cells. As strictly nonbiological entities, "they'd have a much harder time interacting with modern life," Rasmussen said. "They'd be much less of an environmental or health hazard."

Second was their controllability: since they were designed to be programmable, the scientists ought to be able to coax the protocells to perform a larger range of tasks than was possible using ordinary cells and conventional biological engineering techniques. Suitably programmed protocells could unpollute the environment. They could act as "living pharmaceuticals," adapting themselves to a given individual's changing medical needs. They could produce new fuels, chemicals, structures, materials, and technologies.

Finally, because of the commercial value of those activities, they might even—unlike most other research projects financed by the government—make a profit.

AS IT TURNED OUT, money for such a far-fetched project was relatively easy to come by in Europe—provided that you and your organization were based there. Plus, office space for part of the effort was available for free in Venice, as Norman

Packard learned while winding up his previous career in Santa Fe.

Norman was at this stage well into what might be called his Third Major Career Cycle (there were also smaller epicycles). Packard, who happened to be a cousin of David Packard, cofounder of Hewlett-Packard, had started out as a fairly conventional physicist, winning his Ph.D. at the University of California, Santa Cruz, in the late 1970s, after which he pursued a course of research into the main problem areas of the day: chaos theory, self-organizing systems, artificial life. That was his First Career Cycle. (As an epicycle to which, he and his friend Doyne Farmer designed and built miniaturized computer systems that were able to predict, fairly reliably, where a roulette ball would land after a spin of the disk. Never averse to making money with physics, Packard, together with Farmer and Mark Bedau, all of whom had been friends since their undergrad days at Reed College, secreted these devices on themselves and brought them into the casinos of Nevada—until they were busted by the gaming authorities.)

Later, Packard and Farmer founded the Prediction Company, a financial-markets consulting firm in Santa Fe. After several years of successfully modeling, anticipating, and forecasting the allegedly "unpredictable" behavior of the stock market, the business had made small fortunes for both of them. That was Packard's Second Career Cycle. At that point, "it seemed like the right time to try and break loose and pursue some other agendas," he said.

The agenda for his Third Career Cycle was established at a meeting with Rasmussen, Bedau, and their friend John

McCaskill, a theoretical chemist from Sydney, Australia, who by that time had occupied several prestigious academic posts in Germany. The meeting was held at a villa in the Italian resort town of Cannobio, on Lago Maggiore, at the foot of the Swiss Alps. It was about a hundred miles from Lake Geneva, where Mary Shelley had gotten the idea for and started writing *Frankenstein*. The villa was owned by the family of Packard's wife, Grazia.

By the time of this gathering in the summer of 2002, these four researchers—Packard, Rasmussen, McCaskill, and Bedau—were old friends, and had closely shared scientific interests, orientations, and ambitions. Ever since a similar meeting two years before at Ghost Ranch in New Mexico (the former home of artist Georgia O'Keeffe, which was later made into a conference center), the four of them had become increasingly fixated on Steen Rasmussen's protocell design plan. Now, after a week's worth of discussions in Cannobio, the band of brothers decided that the time had come to implement Steen's design. They laid out an organizational plan, a timetable for action, and an informal division of labor, and then they mutually pledged themselves to actually building a protocell.

When they arrived in Cannobio, they were four investigators in search of a project. By the time they left they were the Four Protocell Musketeers. Then they disbanded, each to carry out his allotted part of their overall vision.

First, John McCaskill would apply to the European Union for a grant to establish an international consortium to be known as PACE, an acronym for Programmable Artificial Cell Evolution. Its primary and ultimate objective would be to exploit the programmability of protocells if and when they were brought into existence, but in the interim PACE would function as an umbrella organization that would provide guidance and research money to several European member institutions.

In the fall of 2003, PACE was funded by the European Commission (the executive branch of the European Union) in the amount of 6.6 million euros (about $8.6 million). Switzerland and Lithuania, which were not members of the EU but were nevertheless interested in the PACE project, kicked in with additional money. Soon John McCaskill acquired lab space at an outlying branch of the Fraunhofer Institute, a legendary organization with research centers spread out all over Germany, and gathered together a staff of eleven.

Second, Steen Rasmussen would apply to the Los Alamos lab for funding. In October 2004, his request was granted, and Los Alamos National Laboratory (LANL) appropriated $4.5 million for the Protocell Assembly project.

Third, Packard and Bedau would set up and run a profit-making company that would finance future protocell research in the event that sufficient funding was not obtained through other channels. The company that emerged was ProtoLife s.r.l.,[1] founded with $600,000 in seed money from private "angel investors," the major contributor being Norman Packard himself. The firm would be located at Parco Vega in the city of Marghera, across the lagoon from Venice, in the industrial park the Veneto Regional Government had created as a high-tech research heaven (and named it after a star).

Packard, Grazia, and their two children relocated to Venice, while renting out their Santa Fe house, and moved into a top-floor apartment in the tallest residential structure on the Grand Canal, the Palazzo Contarini degli Scrigni. Bedau, meanwhile, took a year's leave from Reed College, where he was a philosophy professor, and similarly moved himself and his family to Venice.

Fourth and finally, the protocell four would establish a "Center for Living Technology" as a conference, coordinating, training, and educational center—a place where the project's scientists could plot protocell construction strategies and from which the world at large could be informed of the "living technology" that a race of protocells would make possible. A living technology, they theorized, would be one that conferred many of the benefits of living systems—autonomy, robustness, adaptability to environmental changes, self-repairability, and so on—although it would be based not on machines but on programmable chemical cells.

The European Center for Living Technology (ECLT) was born in December 2004 at the Palazzo Giovanelli in Venice. The Palazzo Giovanelli was a three-story pink Gothic mansion that had been built in the first half of the fifteenth century. The place was on the Grand Canal, and the interior was fitted out with marble stairways, gilt-edged doors, mirrors, chandeliers, ballrooms with coffered ceilings, arched windows, and, on the top floor, a stained-glass skylight. The center had been established at Giovanelli not because of its opulence, but because the University of Venice, which had rights to the building, gave ECLT the run of the entire third floor for free.

Suddenly the Four Protocell Musketeers were awash in almost $14 million and flush with laboratories, equipment, simulation facilities, and scientific staff. All they needed now was to build their little organism.

BY MID-2005, all four protocell operations were up and running. At Los Alamos, Rasmussen's group had built containers. "We have pieces of the metabolism working too," Rasmussen said. "What we're working on right now is to integrate the metabolism with the gene."

In Venice, ProtoLife was trying to get its vesicles to perform functions that would create an initial earnings stream for the company.

"Long before ProtoLife is creating any revenue from freestanding, autonomous, self-reproducing artificial cells," Norman Packard said, "it will be producing revenue from much more primitive systems that will be engineered using some of the same tools that we're developing to take us along the path to the artificial cell."

For example, using a variety of hydrophobic-hydrophilic building blocks ("We have nineteen different kinds in our refrigerator," Packard said), ProtoLife chemists were custom-designing vesicles that could be used as targeted drug-delivery vehicles. These would be valuable commodities, especially if they had the ability to evade the human immune system. (Such structures were known within the drug industry as "stealth vesicles.")

None of the Protocell Four, Packard said, believed that their project would create freestanding, living, autonomous cells within the four-year duration of their EU and American funding, "but everybody hopes that it will produce some version of an artificial cell that can exist on technological life support."

Meaning what?

"You'd better ask John McCaskill," Packard said.

John McCaskill's lab was in the German town of Sankt Augustin, a few miles outside of Bonn. It was in a parkland setting, boasted its own castle (the Schloss Birlinghoven), and was a place where, in the early morning, it was not unusual to see people taking their exercise on horseback.

"Norman's lab is just getting started," McCaskill said in Sankt Augustin. "We've been doing this for a while."

That was evident from the quantity of their machinery and instrumentation. One room contained three generations of reconfigurable computers. There were chemical etching machines; supplies of dangerous chemicals; a laser lab; and, most important, a room where the scientists made their own three-dimensional, microfluidic chips.

Microfluidic chips were like computer chips except that fluids coursed through them along with the more usual electric current.

"A microfluidic system is a system which contains flows of material at the microscale, sub-nanometer scale," McCaskill said. "An example would be current tools for chemical diagnostics where one performs tests on chemicals in parallel by flowing the chemicals down these small channels and observing the reactions that take place in them."

A finely controlled fluidic environment was to be the protocell's birthplace, cradle, and life-support system.

"It's like an iron lung for artificial cells," McCaskill said.

Under the laser microscope that was necessary to observe

it, McCaskill's microfluidic system looked like an ordinary computer chip: wires and channels running in a maze along precise, geometrical pathways. The important thing about those channels, however, was the variety of chemicals that could be made to flow through them at extremely slow and precise rates, while the microelectrodes could be made to deliver comparably minute electrical impulses, all of it under real-time direct observation and full computer control. Until it could exist on its own, that precisely controlled flow of chemical nutrients would constitute the cell's feeding system—a microscopic, mechanochemical womb.

In McCaskill's scenario, something like the first baby protocell, a bare five nanometers across, would be born inside such a microfluidic womb by 2008, and would be kept alive by deft manipulation of inputs and outputs until it could exist on its own.

"The task of reaching a freestanding artificial cell," he said, "then just becomes one of successively withdrawing the life support."

Once it had been withdrawn, would the independently existing protocell constitute a genuine life-form? The answer hinged on knowing what life, in essence, really was—a question that scientists and philosophers had contended with for decades, with scant success.

In 1943 Erwin Schrödinger, the quantum physicist, wrote what was destined to become a celebrated and highly influential book devoted to just that question, which he nevertheless discreetly refrained from answering.

Two

Schrödinger

IN FEBRUARY 1943, before the structure of DNA had been discovered, and even before DNA was known to contain the genetic information, the Austrian physicist Erwin Schrödinger gave a set of three public lectures on biology at Trinity College, Dublin. At the age of fifty-six, Schrödinger was a handsome man with sharp, chiseled features and a lively, engaging wit. His lectures, which were given on three successive Friday nights, drew an audience of about four hundred. As he later recalled with no trace of false (or true) modesty, the audience "did not substantially dwindle, though warned at the outset that the subject-matter was a difficult one and that the lectures could not be termed popular."

Such a hit were they that each talk had to be repeated the following Monday in order to accommodate the overflow crowd, which included cabinet ministers, diplomats, socialites, artists, and the president of Ireland, Eamon de Valera. It was as much a social extravaganza as it was a series of scientific talks. Back in the States, *Time* magazine ran a story about the event, making note of Schrödinger's "soft, cheerful speech, his whimsical smile."

In 1944, Schrödinger published a book based on the lectures; like the talks themselves, the book was entitled *What Is Life?* In short order this small work of fewer than a hundred pages became one of the most iconic and influential texts in the annals of twentieth-century biology. It launched a thousand geneticists on their careers, including Maurice Wilkins, James Watson, and Francis Crick, all of whom were swept away by Schrödinger's potent belief that in the near or distant future biology would be reduced to chemistry and physics, thereby dispelling forevermore the long-standing and previously impenetrable "mystery of life."

Life, at that time, was still substantially a mystery. What was its basic nature? How did it get started? What was the driving force that made it go? How precisely did heredity operate in cells? Nobody knew. In 1932, less than a dozen years prior to Schrödinger's talks, Niels Bohr, another physicist, had said: "The existence of life must be considered as an elementary fact that cannot be explained, but must be taken as a starting point in biology." A rather gloomy outlook even for Bohr, who was noted for his occasionally fatalistic pronouncements.

Erwin Schrödinger wanted to challenge the notion that at the core of life was some impalpable excrescence that lay beyond the grasp of science. That was vitalism, which had been a perennial view of life down the centuries. In place of it

Schrödinger wanted to show, by relying on some recent find-ings in biology and physics, that life, albeit something special and indeed a unique phenomenon in nature, was neverthe-less no more mysterious in principle than the inner workings of a windmill, an alarm clock, or an internal combustion engine. Whatever life's puzzles, and they were considerable, they would eventually yield to the steady advance of experi-mental science and rational thought.

That was a fresh perspective, but Schrödinger had always been an independent thinker. He was an only child, born in Vienna in August 1887, to a father who ran a linoleum manu-facturing plant and a mother who was an amateur musician. He was home-schooled for a time, and showed an early dis-trust of authority, remarking, at the age of four: "This says Mama, and that says Aunt. They are both only people. They could just as well say the opposite."

With a mind like that, he could end up only as a scientist or a philosopher, and at length he suffered both fates. Schrödinger entered the University of Vienna and in 1910 re-ceived a doctorate in physics. Even as a lieutenant in the Fortress Artillery during World War I he managed to perform a few experiments just to keep in practice, and he also wrote some physics papers. But he had an equally strong interest in philosophy and often engaged in binge reading in the sub-ject, concentrating on Schopenhauer, the great pessimist, and the Vedanta. He was not without a mystic streak of his own, and wrote miscellaneous gnomic notes to himself such as: "The goal of man is to preserve his Karma," and: "The ego or its separation is an illusion."

While he never gave up his attachment to Hinduism,

physics was decidedly Schrödinger's profession, and he took up a series of teaching positions at European universities. His main contribution to the science was what has become known as the Schrödinger wave equation, which offered a precise mathematical statement of the principles governing wave-particle duality.

Wave-particle duality was the doctrine within quantum physics which held that the fundamental particles of matter could also be regarded as waves. The idea that some of the primal entities of the universe could be regarded equally as either waves or particles was one of the cornerstones of twentieth-century physics. Although the notion made little or no sense intuitively—and even physicists encountered difficulties in trying to explain it on an everyday, commonsense level—wave-particle duality was nevertheless supported by a considerable body of mathematical theory and empirical evidence. An easy lab experiment showed that a stream of electrons, which were classic point-particles, could produce wavelike interference effects, a phenomenon that resulted when the hills and valleys of two out-of-phase waves canceled each other out. So even though electrons were particles, they nevertheless sometimes acted like waves; thus, wave-particle duality.

Between the years 1926 and 1928 Schrödinger produced six papers on wave mechanics, in one of which he put forth his wave equation. In his 1963 *Lectures on Physics*, Richard Feynman summarized the impact of Schrödinger's equation on physicists.

"For many years the internal atomic structure of matter had been a great mystery," Feynman said. "No one had been able to understand why matter held together, why there was chemical binding, and especially how it could be that atoms could be stable ... Schrödinger's discovery of the proper equations of motion for electrons on an atomic scale provided a theory from which atomic phenomena could be calculated quantitatively, accurately, and in detail. In principle, Schrödinger's equation is capable of explaining all atomic phenomena except those involving magnetism and relativity. It explains the energy levels of an atom, and all the facts of chemical binding."

Schrödinger's wave equation won him the 1933 Nobel Prize in physics (which he shared with P.A.M. Dirac, another proponent of wave mechanics). At the time he received it, Schrödinger was teaching physics at Oxford. Later, in 1940, at the personal invitation of Eamon de Valera, who was himself a mathematician, Schrödinger moved to the new Dublin Institute for Advanced Studies, where he became a senior professor and the first director of the School of Theoretical Physics. Soon he turned his attention to genetics.

GENETICS, at that time, was in a state of development that was almost closer to Gregor Mendel than it was to Watson and Crick. When Mendel formulated the laws of inheritance during the mid-1860s, he had no idea what the physical basis was of the hereditary patterns he observed. The term "gene" was not yet in use, and Mendel himself postulated that inheritance was carried by something he called the *Merkmal*, a catchall term translated as "factor" or "determinant." Even in 1909, when the Danish biologist Wilhelm Johannsen coined the word "gene," all he meant by it was a vague something that existed in the reproductive cells and was responsible for passing on heritable traits from parent to offspring.

However, genes had a separate function that was wholly distinct from their role in heredity: they also determined the total set of characteristics of any given organism. It's in this sense that we speak of a "gene for blue eyes," or for brown eyes, or for practically any other physical (or even mental) trait of an individual. The sum total of an organism's genes was responsible for making it the unique organism it was. But as late as the 1940s, what the genes were made of, and how they actually operated in cells, was still largely unknown to biologists.

It was clear that the genes had something to do with the chromosomes, the web of threadlike structures in the nuclei of cells. Through the microscope, scientists could observe the chromosomes splitting apart during cell division, a process that was mesmerizing to watch. They could stain the chromosomes to bring out certain features; they could view them under ultraviolet light; they could X-ray them; subject them to heat and cold; dry them out, centrifuge them, fix them to glass substrates, break them up into smaller pieces. Still, what the chromosomes were made of and exactly how they worked in the hereditary process remained a mystery.

One thing that was clear about them, however, was that the chromosomes were not themselves the genes. This was evident from the fact that there weren't enough chromosomes in the cells of a given organism to fully explain the multiplicity of its traits: human cells, for example, had only forty-six chromosomes, but each person was made up of an uncountable number of individuating characteristics. A gene, therefore, had to be merely a *portion* of a chromosome, which was another way of saying that chromosomes were *collections* of genes.

But what did the genes consist of? For Schrödinger, as for biologists generally, that was one of the big "mysteries of life."

There seemed to be two main possibilities: one was that the genes were proteins, the other was that they were nucleic acids. Phoebus Levine, a biochemist at the Rockefeller Institute (later Rockefeller University), held that the genes weren't likely to be nucleic acids because, like the chromosomes themselves, the nucleic acids were composed of too few types of chemical units (the four lonely bases, adenine, thymine, guanine, and cytosine) to make possible the kind of diversity and complexity seen in biological organisms. Proteins, by contrast, could be composed of up to twenty different amino acids, a number that allowed for a far larger sum total of possible combinations. The genes, then, were more likely to be made up of proteins than of nucleic acids, but the possibility remained that they could be composed of something else entirely.

Schrödinger's interest in these issues had been stimulated a few years earlier when a German physicist friend of his, Paul Ewald, sent him a copy of what has come to be known as "the three-man paper." It was called "The Nature of Genetic Mutations and the Structure of the Gene," and had been written by Max Delbrück and two coauthors in 1935.[1] The general concept proposed by the three scientists was that genes were composed of relatively small numbers of atoms, and that when the chemical makeup of those atomic structures changed, mutations occurred.

"We leave open the question whether the single gene is a polymeric entity that arises by the repetition of identical atomic structures," they wrote, "or whether such periodicity is absent; and whether individual genes are separate atomic assemblies or largely autonomous parts of a large structure, i.e., whether a chromosome contains a row of separate genes like a string of pearls, or [is] a physico-chemical continuum."

That was a lot of questions to leave open, but they were catnip to Erwin Schrödinger, to whom the idea that the genes were composed of small groups of atoms—each group acting as a functional unit—constituted a highly unusual, even unique, situation. Normally, events in the macroworld, the world of everyday experience, were a function of large-scale flows of atoms and molecules, vast torrents of atoms measured out in droplets, gallons, or tons of matter. In the universe of inanimate objects, rarely did anything of importance hinge on the disposition of one, two, or some other very small number of atoms.

"In biology we are faced with an entirely different situation," Schrödinger noted. "The arrangements of atoms in the most vital part of an organism differ in a fundamental way from all those arrangements of atoms which physicists and chemists have hitherto made the object of their experimental and theoretical research. The situation is unprecedented, it is unknown anywhere else except in living matter."

That was one way in which life and its processes were unique in nature. But there was another. As a quantum physicist, Schrödinger knew better than anyone else that tiny atomic structures were constantly being pummeled, poked, and batted around by the vibrating sea of molecules they were surrounded by. This was the lesson of Brownian motion, which was visible evidence of the forces at work at the molecular level. But how could a gene—a collection of atoms—remain stable amid such turmoil, especially across the course of several generations?

Schrödinger expressed the difficulty in a vivid image, using the *Habsburger Lippe* as an illustration. That trait, a drooping lower lip, manifested itself throughout the Hapsburg dynasty from the sixteenth century onward. "How are we to understand that [the Hapsburg lip gene] has remained unperturbed by the disordering tendency of the heat motion for centuries?" he asked.

More generally, if the genes were precise molecular-scale collections of atoms, and therefore subject to all the disruptions of heat motion, then how did biological traits persist across time and get passed down successfully from generation to generation? It didn't seem possible.

But there was a third way in which life and its processes were special and differed from what occurred in nature's wider realm of inert matter. The most characteristic feature of life, in Schrödinger's view, was that it was a process of constant motion and activity. A thing was considered to be alive, he said, when it was "'doing something,' moving, exchanging material with its environment, and so forth, and that for a much longer period than we would expect an inanimate piece of matter to 'keep going' under similar circumstances."

What was unique about the situation was that such self-perpetuating activities seemed to escape the all-embracing and ironclad second law of thermodynamics according to which events ought to run out of steam, taper off, descend into chaos, and finally cease altogether as they progressively radiated their waste heat to the external environment.

"When a system that is not alive is isolated or placed in a uniform environment, all motion usually comes to a standstill very soon as a result of various kinds of friction," Schrödinger said. "The whole system fades away into a dead, inert lump of matter. A permanent state is reached, in which no observable events occur."

In physics, that was known as a state of thermodynamic equilibrium, or what was the same thing, a state of maximum entropy. Living things, however, were to all appearances absolutely immune to all such influences, at least while they remained alive.

And so, for that matter, did species themselves over the course of evolution, a process in which organisms did not run down, give out, or blend together into a state of uniform dullness, but on the contrary became more ordered, better adapted, and more individuated and diversified as time went on. Life exhibited no propensity for any descents into chaos. Just the opposite: living things were jewel-like oases of harmony, growth, and self-regulating organization in an otherwise barren universe.

"How does the living organism avoid decay?" Schrödinger asked.

Those three problems—the unique dependence of life and its processes on tiny groupings of atoms within the gene; the stability of the gene across time; and the apparent exemption of living organisms from one of the most basic laws of nature—these were the "mysteries of life" that most concerned Erwin Schrödinger.

IN HIS LECTURE SERIES and his book, Schrödinger disposed of each puzzle in succession.

With respect to life's alleged exemption from the laws of thermodynamics, Schrödinger argued that, despite appearances, life did not in fact have any thermodynamics-defying qualities whatsoever. Organisms released heat energy into the environment and, once gone, that energy was unavailable to do further work. The loss of that energy increased the sum total of entropy (disorder) in the universe, exactly as the second law of thermodynamics dictated. The way living things "kept going," Schrödinger said, was by withdrawing what he called "negative entropy" from the environment.

Negative entropy, however, was just his whimsical private term for order.

"The device by which an organism maintains itself stationary at a fairly high level of orderliness . . . really consists in continually sucking orderliness from its environment," Schrödinger said. "In the case of higher animals we know the kind of orderliness they feed upon well enough, viz., the extremely well-ordered state of matter in more or less complicated organic compounds, which serve them as foodstuffs. After utilizing it they return it in a very much degraded form."

The persistence of life was therefore a case of getting order from order: ordered organisms consuming ordered nutrients. Living entities then converted some of that order into disorder, which was perfectly consistent with the second law of thermodynamics. End of mystery.

The second of life's enigmas as defined by Schrödinger—the question of how the gene's inner molecular structures could be stable across time—easily dissolved in the face of Schrödinger's conjecture as to what the gene consisted of. It was not just a bunch of atoms linked together like a string of pearls, he said; rather, it was *an aperiodic solid*—an object with a fixed but nonrepeating structure.

"The gene is most certainly not just a homogeneous drop of liquid. It is probably a large protein molecule, in which every atom, every radical, every heterocyclic ring plays an individual role."

The Hapsburg lip could remain a persistent trait across the centuries, then, because of the way in which the corresponding gene's atoms were arranged inside the chromosome. Those atoms were bonded together in a relatively rigid molecular framework, and were not floating around chaotically inside their sausagelike containers. The chromosomes might look insubstantial and filmy under the microscope, but at the atomic level their structures were as ordered and fixed as the Pyramids.

Still, the final problem remained: How could the small numbers of atoms that made up the gene play their pivotal roles in directing the processes of life? How could tiny atomic occurrences and isolated molecular rearrangements at the bottom of matter yield such gigantic effects on the macroscale?

Schrödinger's answer to this would be his claim to fame in the life sciences. It introduced one of the most characteristic and fundamental concepts of what was soon to become the science of molecular biology: the assertion that *the gene was a message written in code*. The genes, he said, "contain in some kind of code-script the entire pattern of the individual's future development and of its functioning in the mature state. Every complete set of chromosomes contains the full code."

That was prophetic, for there, in essence, was the kernel notion of molecular biology: the idea that the gene embodies *information*, stored and expressed in the form of specific groupings of atoms. The idea of a message system inside the chromosomes had been hinted at by some earlier theorists, but Schrödinger was the first to formulate the idea so crisply and clearly, and to give it a precise physical interpretation as a strictly biochemical phenomenon. Even today, the idea that the basis of life is a code, a *language*, is a radical, almost fantastic notion.

Schrödinger did not specify what language the code-script was written in, how it originated, or how it operated, except to say that it was physically embodied in an aperiodic solid in which groups of precisely arranged atoms coded for particular physical hereditary traits. The gene, he predicted, would be found to consist of "an unusually large molecule which has to be a masterpiece of highly differentiated order," one in which "every atom plays an individual role."

Further, "the number of atoms in such a structure need not be very large to produce an almost unlimited number of possible arrangements. For illustration, think of the Morse code . . . With the molecular picture of the gene it is no longer inconceivable that the miniature code should precisely correspond with a highly complicated and specific plan of development and should somehow contain the means to put it into operation."

Finally, that molecular message, in whatever language it was written in, was sufficiently comprehensive and precise as to be able to determine the full flowering and ultimate development of the fertilized egg. The arrangements of its coded elements determined, he said, "whether the egg would develop, under suitable conditions, into a black cock or a speckled hen, into a fly or a maize plant, a rhododendron, a beetle, a mouse or a woman."

That was as far as Schrödinger went in interpreting the molecular directives that lay at the base of heredity and that determined the structure and traits of any given organism. Further progress would be made along these lines, he was sure, but it would not occur anytime "in the near future."

WHATEVER THE CORRECTNESS of its arguments—about which there would be some debate—*What Is Life?* struck a nerve among scientists, particularly physicists, many of whom had previously viewed biology as a second-class and rather dubious branch of knowledge. Now, life and its processes were no longer as unknown as the far side of the moon, for here was a physicist, a modern quantum physicist no less, telling the world that life could be explained in the same terms, and by the same laws of physics and chemistry, as those that pertained to everything else in nature. Living entities constituted a special sort of natural phenomenon not because they were unexplainable, which they were not, but because their activities were determined by such a paltry number of atoms in relation to the overall size of the organism. Exceptional as it was, life was as rule-governed, law-abiding, and understandable as anything in the material universe. There was no magic or mystery about it.

Schrödinger's biographer, Walter Moore, said of *What Is Life?*: "There is no other instance in the history of science in which a short semipopular book catalyzed the future development of a great field of research." The book was translated into seven languages, went on to sell more than 100,000 copies, and became a modern classic of science literature.

Each of the three scientists—James Watson, Francis Crick, and Maurice Wilkins—who in 1962 would share the Nobel Prize for having deciphered the structure of DNA, revealing it to be the aperiodic solid Schrödinger had predicted, credited Schrödinger's seminal book with having been a motivating force in turning them toward biology and genetics.

Crick said of *What Is Life?*: "It suggested that biological problems could be *thought* about, in physical terms—and thus it gave the impression that exciting things in this field were not far off."

Watson, for his part, said: "From the moment I read Schrödinger's *What Is Life?* I became polarized towards finding out the secret of the gene."

But for all its clarity, brevity, and foresightedness, Schrödinger's book had some shortcomings. One was the fact that it did not answer, or even attempt to answer, the question of the title, *What Is Life?* Was the title in truth no more than a come-on, a clever marketing ploy? Or was the answer too mysterious even for Schrödinger? He never said.

The book's other flaw was that, practically at the very moment he was telling his audience in 1943 that the gene was "probably a large protein molecule," three American scientists were finishing up a long series of experiments in which they definitively established that in fact it was not.

Three

Unlocking the Three
Secrets of Life

SCHRÖDINGER HAD PREDICTED in 1943 that no detailed information about "the functioning of the genetical mechanism" would emerge in the near future. As it happened, progress toward that end occurred in three distinct incremental stages, each of which nevertheless embodied a major discovery about the gene, which indeed gave up its secrets slowly. The first disclosure, announced to the world in 1944, was that genes were nucleic acids and not, as many biologists had thought, proteins. With that finding, scientists knew for the first time the composition of the basic entities that controlled the heredity and development of life on earth.

The gene's second secret was the structure of the nucleic

acid molecule itself, DNA. Famously, Watson and Crick brought it to light in 1953 when they proposed that the long strands of DNA corkscrewed around each other artistically in the form of a double helix. That was a key revelation because, for one thing, it provided a physical basis for how heredity worked.

But since the gene had a double function, so too did DNA. In addition to storing the heritable information, the molecule also controlled the processes of biological growth, development, and morphology: it guided the construction of cells, bodily structures, and whole organisms. The information for building those components was stored in the form of a code (Schrödinger's "code-script"), and breaking it was the third riddle that confronted scientists. When a little-known biochemist at the National Institutes of Health did so in 1966, the feat enabled scientists to literally read the language of life.

Each of these fundamental advances centered upon DNA, the iconic molecule of all history. It's the Statue of Liberty, the Venus de Milo of molecules, the one chemical entity the whole world thinks it knows the structure of. Moreover, plenty of educated people imagine that they also know perfectly well what it is, what it does, and how it works. On the other hand, some of those who ought to know better fall flat on their face when it comes to presenting even the most basic facts about the subject. In 1996, for example, an internationally acclaimed science writer and host of two television documentaries about science and technology wrote in his latest book that "it was through the use of [X-ray diffraction] that in 1952 Francis Crick and James Watson were able to confirm the three-dimensional structure of a molecule of protein . . . Because of the discovery of DNA, science is already well on the way to the Biological Revolution."

As it happens, each of those three claims—that DNA is a protein, that Watson and Crick discovered it, and that they did so in 1952—is false.

The "unknown" story of DNA begins with Johann Friedrich Miescher, a Swiss biochemist who discovered the substance in 1869 in purified pus isolates taken from discarded hospital bandages.

Miescher had been particularly interested in the chemistry of the cell nucleus, a subject whose study would require a dependable supply of nuclei. This was well before the age of mail-order delivery of any conceivable variety of biological specimen, and Miescher was forced to create his own private cell nucleus bank. That was by no means a routine operation, however, and he had to trailblaze a path of his own devising in order to amass a suitable cache of raw materials.

Miescher focused on leukocytes, or white blood cells, because their nuclei were exceptionally large and sometimes even appeared in the microscope as double. Still, how did you extract a bunch of nuclei from a mass of leukocytes? A cell was composed of an outer membrane, an inner cytoplasm, and a nucleus, among other things. Extracting the nucleus was not something to be done mechanically, with tweezers (although it could be done that way today), but was rather a process to be performed in bulk, chemically.

Leukocytes are the main ingredient of pus, and so Miescher decided that to harvest a lot of leukocytes he needed a ready supply of pus. He gathered it by collecting used surgical bandages from a nearby hospital, washing the bandages with a dilute sodium sulfate solution and then filtering the residue to remove tissue fibers and other impurities. Since laboratory centrifuges did not yet exist, he let the end product stand for a while so that the cells would settle out to the bottom of the beaker. To extract the nuclei from the cells, Miescher bathed them in warm alcohol to dissolve the outer cell membrane, and used the enzyme pepsin, which he had derived from a pig's stomach, to digest away the cytoplasm that still surrounded the nucleus. The nuclei that remained after this procedure formed a dull grayish gloop.

Analyzing it chemically, he found that it was acidic and that it consisted of 14 percent nitrogen, 3 percent phosphorus, and 2 percent sulfur. Because of its high phosphorus content and its resistance to digestion by pepsin, which catalyzed the breakdown of proteins, Miescher concluded that the substance in question was not a protein but something else.

Since it had come from the nucleus, he called the stuff "nuclein." It was in fact an impure form of DNA, and it was the first known isolation of the substance.

Later, in 1870, Miescher found a better source of nuclein in salmon taken from the Rhine, a river well-known for the fish. Sperm cells, he knew, had large nuclei, and those of salmon were exceptionally fat and distinct. From them he extracted and isolated a more purified form of nuclein, which he called "pure nuclein." Miescher published his results in two papers: "On the Chemical Composition of the Pus Cells," in 1871, and "The Spermatozoon of Some Vertebrate Animals: A Contribution to Histochemistry," in 1874.

In 1879 Albrecht Kossel, a biochemist colleague of Miescher's, analyzed pure nuclein and found that it contained four nitrogen-rich compounds which he christened *adenine*, *thymine, guanine*, and *cytosine*. We now recognize these as the four bases of the DNA molecule, canonically, A, T, G, and C. In 1889, finally, a student of Miescher's, Richard Altmann, renamed the substance "nucleic acid," a name that stands to this day.

In 1874 Miescher had ventured a prescient speculation about the significance of the chemical compound he had discovered: "If one wants to assume that a single substance . . . is the specific cause of fertilization then one should undoubtedly first of all think of nuclein."

OTHER THAN Miescher, though, few scientists linked nuclein with heredity until 1943, the year in which three researchers working at the Hospital of the Rockefeller Institute for Medical Research, Oswald Avery, Colin MacLeod, and Maclyn McCarty, showed, after some thirteen years of work with two strains of pneumonia bacteria, that the genetic information was indeed embodied in nucleic acid. That, at the time, practically bordered on heresy. In biology, the current dogma had it that the carriers of the hereditary data were protein molecules. The proteins, after all, were composed of twenty different amino acids, whereas the nucleic acids were merely boring, monotonous macromolecules that had only the four puny A, T, G, and C components to work with; they therefore seemed totally unsuited to the task of carrying the volume of information that was packed inside the genes. This meant that the genes almost *had* to be proteins.

The research that would ultimately defeat that orthodoxy had its beginnings in a biologically alarming phenomenon: the apparent spontaneous resurrection of dead pneumonia bacteria. In 1928 a British army doctor and researcher by the name of Frederick Griffith, who was hoping to develop a vaccine against the disease, stumbled upon a mysterious event he called "the transformation of pneumococcal types."

There were at least two different strains of pneumococci bacteria. One strain was encapsulated by a smooth carbohydrate coating and was therefore known as *Type S* (for smooth). Type S pneumococci were also highly virulent, and when injected into laboratory mice produced infection and, routinely, death. The other strain was nonencapsulated, had a rough exterior when viewed on the surface of a culture plate, and was known as *Type R* (for rough). Type R pneumococci were noninfective and by themselves were incapable of causing fatal pneumonia. In essence, they were harmless.

With that as the setup, Griffith injected laboratory mice with a small amount of *living* Type R cells (the harmless ones) together with a large amount of *heat-killed* Type S cells (the lethal variety), in hopes that this mix would offer some protective value against pneumonia. As if! Instead, what happened was that the mice inoculated with the mixture keeled over and died just as if they had been injected with the original live S cells. Even greater was his surprise when blood taken from the dead animals was found to contain *living Type S cells*—the lethal strain! It was as if the heat-deactivated lethal bacteria had somehow miraculously resurrected themselves and then proceeded to kill off mice.

Outside of horror fiction, spontaneous resurrection did not seem likely. But the only other obvious possibility was that some element in the dead S cells had migrated over to the live R cells, invaded them, and caused them to acquire a smooth coat and become virulent. That's what must have happened, Griffith concluded: some unknown factor in the dead S cells had caused a transformation in the living R cells, converting them into cells of Type S.

A gene could effect such a transformation. The question was, what was the composition of the gene responsible for the metamorphosis? That was the situation facing Avery, MacLeod, and McCarty when they started working on the problem in the 1930s.

All three men held medical degrees but had decided to fight disease by doing pure research as opposed to practicing medicine. They began by repeating Griffith's work, which had also been replicated by several other researchers independently. Then they went a step further, showing that under the right conditions the original transformation was reversible, meaning that an S cell could be reconverted into an R cell. Second, they showed that the cell types retained their newly acquired characteristics from generation to generation, which meant that a true genetic modification had occurred. Once having been transformed, however, the new cell lines never spontaneously changed back to their former type, meaning that the genetic change in question was permanent (unless a further transformation was experimentally induced). Plainly, some mystery substance in the cells was acting as a gene.

Over a period of about ten years, Avery and his colleagues tried to identify the material in question. Essentially, their procedure followed the Sherlock Holmes principle of ruling out all other possible explanations and being left with only one alternative, which had to be the truth, no matter how unlikely it might have seemed to begin with. In simple terms, what Avery and company did was to isolate and purify what they regarded as the "active principle" of the transformation. They did this by stripping the relevant cells of proteins, impurities, and random contaminants. The "active principle," whatever it was, remained, for when cells transformed by it were injected into mice, the animals died right on schedule.

They then performed some basic chemical tests on the substance, finding that a standard "reaction for desoxyribonucleic acid [DNA] is strongly positive." They chemically analyzed the compound and found that it was composed of carbon, hydrogen, nitrogen, and phosphorus in ratios known to be characteristic of DNA. When they subjected the material to enzymes known to destroy DNA (for example, rabbit bone phosphatase and swine kidney phosphatase), the stuff suddenly lost its previous ability to transform cells. When they placed purified samples of it in electrical fields and in an analytical ultracentrifuge, the substance behaved exactly as DNA did in those conditions.

What else could the substance be, then, but DNA?

At the end of all their various studies and tests, and having ruled out all other reasonable alternatives, the three scientists decided that that's what it was. They wrote up a twenty-page paper, "Studies on the Chemical Nature of the Substance Inducing Transformation of Pneumococcal Types," and in

November 1943 submitted it to the *Journal of Experimental Medicine,* which published it in February 1944. In it, after vast tracts of dense, jargon-laden, data-filled prose, tables, and charts, the authors rather circumspectly concluded that "the evidence presented supports the belief that a nucleic acid of the desoxyribose type is the fundamental unit of the trans-forming principle."

The genes, in other words, were DNA.

DESPITE THE MAGNITUDE of the discovery it reported, the Avery, MacLeod, and McCarty piece received little attention at the time it was published, and some of the attention it did receive was actively hostile.[1] After all, geneticists had been as-suring one another for years that genes were proteins. To be suddenly informed that they weren't went very much against the grain of some—particularly Alfred Mirsky, one of Avery's colleagues at Rockefeller, who proceeded to hold out against the idea for several years. On the other hand, that was often the way of discoveries that overturned conventional wisdom. As James Watson, in his typically charming fashion, later said of the protein enthusiasts: "Many were cantankerous fools who unfailingly backed the wrong horses . . . In contrast to the popular conception supported by newspapers and moth-ers of scientists, a goodly number of scientists are not only narrow-minded and dull, but also just stupid."

Much later, Avery, MacLeod, and McCarty's paper would receive its proper due. Still, it took ten years from their proof that the genes were DNA until the time Watson and Crick re-solved the structure of the molecule in the spring of 1953.

Because it is the core entity of life, controlling heredity, growth, and practically every other biologically significant function or factor, DNA's significance cannot be exaggerated. DNA is to life what atoms are to matter.

As has by now passed into legend, Watson and Crick's discovery of DNA's structure rested in relatively equal parts upon a physical model of the molecule they constructed based on the known facts of its chemical constituents and on a series of X-ray diffraction photographs of the molecule taken by the physicist-turned-biologist Maurice Wilkins and by the crystallographer Rosalind Franklin. The double-helical shape that Watson and Crick postulated was a function of chemical bonding angles, dispersion forces, and ionic interactions (plus other determinants) acting on the molecule's component parts. Specifically, they theorized that the subunits that made up the molecule's backbone did not bond to each other in a straight line. Rather, each was set at an angle of 36 degrees relative to the next, so that the structure made a complete 360-degree turn after ten successive repeats of the subunits. That explained why the molecule was helical.

Watson had been impressed by the stylish fashion in which Linus Pauling had written his scientific papers—"The language was dazzling and full of rhetorical tricks"—and decided to write up his and Crick's findings in the same dashing manner. Mostly, he failed. Nevertheless, what resulted was one of the most historic papers in all of science, a one-page, nine-hundred-word "Letter" to the British science journal *Nature*, published on April 25, 1953. It began: "We wish to suggest a structure for the salt of deoxyribose nucleic acid (D.N.A.). This structure has novel features which are of considerable biological interest." The authors went on to explain their rationale for the structure, and they accounted for the way its dual sugar-phosphate backbones were held together by base pairs that lay in a plane perpendicular to the axis of the fiber.

Watson and Crick had laid bare the construction of the molecule that was at the core of all life on earth.[2] Anyone who has read Watson's version of the story in his book *The Double Helix* will recognize the claim that when they first realized that their structure was correct, Crick walked into the Eagle, a favorite pub of theirs on Bene't Street in Cambridge, and bragged to "everyone within hearing distance that we had found the secret of life."

An understandable boast, but not the whole truth, as is clear from their paper's famous closing line: "It has not escaped our notice that the specific pairing we have postulated immediately suggests a possible copying mechanism for the genetic material." What they had found was the secret of *heredity*, for they had indeed unveiled the physical shape of the biological structure by which genetic sequences were copied and then reproduced in daughter cells.[3]

But there was more to life than heredity, and DNA had more of a role in cells than merely making copies of itself. Indeed, replication was strictly ancillary to the primary purpose of nucleic acids in cells, which was to control the production of the proteins that constituted the physical corpus of any given living thing.

Proteins were composed of amino acids, and a specific sequence of DNA bases held the instructions for stringing together the right combinations of amino acids that produced a given protein. But as Erwin Schrödinger had suggested ten years earlier, those instructions were written in code; in particular they were expressed in a chemical cipher in which a given triplet of bases stood for a distinct amino acid.

When Watson and Crick supplied the DNA molecule's structure, the key to the genetic code was still unknown. Their famed paper does not contain the words "amino acid," "protein," or "genetic code." Another of life's secrets, therefore, lay inside DNA's molecular coding system. Finding the key to it would be like unearthing the Rosetta stone of living things.

IN THE LATE 1950S, an obscure researcher at a United States medical research facility took on the task.

This was Marshall Nirenberg, a biochemist at the National Institutes of Health in Bethesda, Maryland. Nirenberg had been born in New York City but developed rheumatic fever as a boy, and on the theory that a semitropical climate might be curative, the family moved to Orlando, Florida. In 1957 Nirenberg got a Ph.D. in biochemistry from the University of Michigan and immediately accepted a position at the NIH.

Two years later he began to investigate the processes by which DNA directed the building of proteins. By that time scientists had realized that the DNA molecule was basically an information storage complex and was not itself directly involved in protein synthesis. Instead, DNA first transferred its information to an intermediate molecule, RNA.

RNA had been discovered around 1910 by Phoebus Levene,

a Rockefeller Institute chemist. RNA was similar to DNA in that it was a large, long macromolecule composed of a sugar-phosphate backbone and four bases, or repeating subunits. But RNA differed from DNA in several respects. For one thing, whereas DNA was a double-stranded molecule that replicated, RNA was a single-stranded molecule that did not. For another, while they shared three bases (adenine, guanine, and cytosine), they differed in one: DNA's thymine was replaced in RNA by uracil. Third, whereas DNA was DNA, there were several different types of RNA: messenger RNA (mRNA), transfer RNA (tRNA), and ribosomal RNA (rRNA), among others. Finally, there was lots more RNA in cells than there was DNA; in fact, RNA was five to ten times more abundant than DNA in an average cell.

The cells built proteins in a stepwise process. First, the DNA molecule zipped apart down the middle and transferred its information to RNA, specifically to messenger RNA. Second, the mRNA exited the nucleus and entered the cell's cytoplasm. There, its coded message was read by tiny cytoplasmic organelles called ribosomes which then proceeded to string amino acids together of the type and in the order prescribed by the mRNA bases.

At NIH, Marshall Nirenberg speculated that if he placed into the cytoplasm a known type of RNA molecule he might be able to stimulate the production of a specific type of amino acid. If such a scheme worked, it would be a key to the genetic code: RNA sequence X would code for amino acid Y, and so on. The underlying principle was one of covariation: it would be something like throwing circuit breakers in an elec-

trical panel one by one and then finding out by empirical observation which particular electrical outlet each controlled.

He took *E. coli* bacteria as his microbe of choice. *E. coli*
are the workhorse organisms of molecular biology: they are
common and well-studied intestinal bacteria, they reproduce
reliably every twenty to sixty minutes depending on their growth
conditions, and therefore constitute a dependable and cheap
supply of microbial guinea pigs. Further, they have the added
advantage of being prokaryotic cells, meaning that they contain no cell nucleus: their DNA, RNA, and everything else
floats around openly in the cytoplasm. This allows scientists
to convert them easily into one of the strangest phenomena of
all life, the so-called "cell-free system," which is essentially
a large mass of cytoplasm and its associated contents, but
without any enclosing cell wall. A cell-free system was a free-
floating monster cell unencumbered by a membrane.

In his experiments, Nirenberg created a cell-free system
by grinding a mass of *E. coli* bacteria with alumina, a common abrasive, extracting the liquid (also known as cell sap),
and then centrifuging it to separate the ground-up cell walls
from everything else. Individual cells now no longer existed,
only their inner contents, which formed a sort of mass cytoplasmic soup, a yellowish, not very viscous, opaque fluid.

The shock is that this cytoplasmic soup nevertheless acted
as if it were still a living entity. This dynamic blob was like
a horror-movie supercell in that since it contained all the cellular machinery—DNA, RNA, ribosomes, and enzymes—
necessary to make proteins, it could do practically everything
the original cells did except for reproducing itself. That was

why, in Nirenberg's view, a cell-free system was not alive: "It's prepared from living things," he said much later. "It can synthesize proteins. It can maybe even synthesize DNA. But it can't reproduce, so it's not living."

Whether reproduction is a necessary condition for life is a matter for some debate; indeed, it is one of the central issues that must be decided in any attempt to answer the question "What is life?" But alive or not, when Nirenberg started experimenting with his cell-free systems, he found that by adding the enzyme DNAase he could shut down protein synthesis altogether. On the other hand, he could start the process up again by adding certain mRNA extracts to the mix—both of which moves told him that the cell-free system was working the way it was supposed to.

Now for the crucial step, which was to introduce into this semiliving, semidead biochemical production apparatus an RNA molecule of known identity, and then to see what the response was, if any. It would be like switching a given circuit breaker to the ON position and hoping that a light turned on someplace in the house.

Other scientists, Nirenberg was aware, had recently synthesized strings of RNA composed of just a single, repeating base, for example, a triplet of uracils—UUU—which was called "poly-U." That would be the input molecule. What would the output be? What would the cell-free system make in response?

Working in Nirenberg's lab was a German postdoc by the name of Heinrich J. Matthaei. Late one night in May 1961, Matthaei added a quantity of poly-U to their cell-free system. This, in retrospect, should have been a hair-raising moment.

More than that, it should be remembered as one of the most fateful steps ever taken in the breaking of the genetic code, for here was a scientist inserting a single, isolated, and known ingredient into a carefully constructed cellular production works. If nothing happened, the experiment would have been a dud. But if he got an identifiable response, he would have found the key to breaking the entire genetic code, right then and there, on the spot.

Which is exactly what occurred. When the cell-free system received its input shot of UUU, it responded by churning out the amino acid phenylalanine.

All at once, the genetic code had been broken: UUU stood for phenylalanine. That solitary and cryptic UUU triplet was the first word in the chemical dictionary of life, and it was the key to deciphering the rest of the genetic code.

Three months later, in August 1961, Nirenberg presented his results publicly for the first time, at the International Congress of Biochemistry, in Moscow. On the plane over, he sat next to another NIH scientist who asked what Nirenberg was up to. "I told him what I was going to present at the meeting, and it seemed to go in one ear and out the other."

Only a scattering of people were in the audience for Nirenberg's talk. Reading from three pages of partly typed, partly scribbled notes, he presented his results in a small amphitheater to a group of about twenty-five. Neither Watson nor Crick, who were both at the conference, attended. One who did was Alfred Tissières, a Swiss biochemist who actually understood the importance of what he was hearing. Tissières spread the word, and Crick invited Nirenberg to give his talk again.

"The second time was to a very large audience," Nirenberg remembered. "There the reception was really remarkable, fantastic. I remember Matt Meselson, who was sitting right up front. I didn't know him at the time, but he was so overjoyed about hearing this stuff that he impulsively jumped up, grabbed my hand, and actually hugged me and congratulated me for doing that. I could have been part of a rock band or something!"

Over the next four years Nirenberg, together with Har Gobind Khorana and Robert Holley (plus a small fleet of postdocs), worked to decipher the remainder of the genetic code.[4] The next word in the genetic lexicon was the cytosine triplet, CCC, which coded for the amino acid proline. Poly-A (the adenine triplet, AAA) coded for lysine, poly-G (guanine, GGG) coded for glycine, and so on.

By 1966 Nirenberg and the others had deciphered all the sixty-four codons, the three-letter words of mRNA for all the twenty different amino acids. Further experiments on different kinds of organisms revealed that the genetic code was universal (or almost) among all the life-forms on earth.

"This had a big philosophic impact on me," Nirenberg said later. "I thought this was really extraordinary, that I'd look out the window and see a tree and maybe a squirrel sitting in the tree, and I'd think that the instructions in the plant and the squirrel are really the same. It had a big emotional impact on me."

And why not? If nature was written in the language of mathematics, life was written in the language of the genetic code—the very code that Nirenberg had been instrumental in breaking. Two years after their work was complete, Nirenberg, Khorana, and Holley won the 1968 Nobel Prize in physiology or medicine for "interpretation of the genetic code and its function in protein synthesis."

Together, these three discoveries—that the genes were DNA, that the structure of the DNA molecule provided the physical basis of heredity, and that the message expressed in DNA could be decoded by means of the correspondence between RNA triplets and amino acids—formed one of the largest conceptual advances in the history of biology. Life and its processes could be *understood*, and they could be understood rationally, in terms of known science, the ordinary laws of physics and chemistry. All of which would take us one step closer to finally answering the question "What is life?"

Four

The Fiftieth-Anniversary
Coronation and Dismissal

SCHRÖDINGER'S PIONEERING LITTLE BOOK had such a lasting impact among scientists that in 1993, fifty years after the original lecture series, a three-day scientific conference titled "What Is Life? The Next Fifty Years" was held at the original scene of the crime, Trinity College, Dublin. Such celestial bodies of the modern biological firmament as Stephen Jay Gould, John Maynard Smith, Manfred Eigen, Stuart Kauffman, and Leslie Orgel, among others, were on hand for the purpose of commemorating Schrödinger's seminal and prophetic achievements. But as would become clear over the following days, the scientists came both to praise Schrödinger and to bury him. (Schrödinger himself, meanwhile, had died in Vienna on January 4, 1961, at the age of seventy-four, of heart disease and atherosclerosis. He was buried in a Catholic churchyard in the village of Alpbach, in the Austrian Tyrol, where he often spent his summers.)

The somewhat schizoid attitude the conference participants manifested toward Schrödinger mirrored the critical reception of his book from the beginning. Indeed, there had been problems even before publication, stemming from a highly puzzling five-page epilogue the author had appended to the main text, an essay entitled "On Determinism and Free Will." Schrödinger had justified its inclusion by saying that "as a reward for the serious trouble I have taken to expound the purely scientific aspect of our problem *sine ira et studio* [without anger or bias], I beg leave to add my own, necessarily subjective, view of its philosophical implications."

Thereupon Schrödinger tackled a classic philosophical conundrum: How could the strictly mechanical functioning of the body according to the ironclad laws of nature possibly be compatible with one's own personal, direct experience of free will? In philosophy, this chestnut fell into the realm of metaphysics, and so it did here. Answering the question, Schrödinger sank deep into Indian mysticism and the Upanishads, advancing the arcane doctrine that "ATHMAN = BRAHMAN" (which meant that the personal self equaled the omnipresent, omniscient, and eternal self), and drawing liberally upon the Vedanta, all in an attempt to make palatable his own private solution to the question: "I in the widest meaning of the word, that is to say, every conscious mind that has ever said or felt 'I'—am the person, if any, who controls the 'motion of the atoms' according to the Laws of Nature." That memorably obscure pronouncement in turn meant, he said, "Hence I am God Almighty," or, put another way, "DEUS FACTUS SUM (I have become God)."

In the wake of his extremely careful and cautious elucidation of the workings of molecular genetics, this was pretty wild stuff. Schrödinger himself admitted as much, acknowledging that for Christians the doctrine was "both blasphemous and lunatic." He himself was no Christian, however, and affirmed this idiosyncratic viewpoint with all the assurance of a transatlantic ocean liner steaming through the Strait of Gibraltar.

The publishers to whom he submitted the book, however, Cahill & Co., of Dublin, were not especially pleased with this outburst. Proper Christians themselves, they refused to print the epilogue, whereupon Schrödinger refused to delete it. He then sent the manuscript to Cambridge University Press, which brought out the complete and unexpurgated text in 1944.

The scientists who paid any attention at all to the mystic epilogue were appalled by its teachings. Hermann J. Muller, a well-known geneticist, said in a 1946 review of the book: "If the collaboration of the physicist in the attack on biological questions finally leads to his concluding that 'I am God Almighty,' and that the ancient Hindus were on the right track after all, his help should become suspect." Others politely averted their eyes from Schrödinger's temporary foray into occultism, viewing it as the physicist's personal philosophic opinion which had to be separated rigorously from his science.

The intervening years were generally kind to *What Is*

Life?, although there were a few notable holdouts to the steady flow of accolades. In 1987, on the centenary of Schrödinger's birth, Linus Pauling reported that "when I first read this book, over forty years ago, I was disappointed. It was, and still is, my opinion that Schrödinger made no contribution to our understanding of life." The biochemist Max Perutz, writing at about the same time, said: "A close study of his book and of the related literature has shown me that what was true in his book was not original, and most of what was original was not known to be true even when the book was written. Moreover, the book ignores some crucial discoveries that were published before it went into print." (Chief among these was Avery, MacLeod, and McCarty's demonstration that the genes were nucleic acids, and not, as Schrödinger had said, proteins.) The molecular biologist Gunther Stent added: "The title was a piece of colossal nerve."

WHEN THE CLUTCH of eminent scientists gathered together for the fiftieth-anniversary celebration in Dublin, though, their remarks gave the impression that biology, rather than advancing, had taken several steps backward in the interim. Whereas Schrödinger had radiated a level of confidence approaching irrational exuberance about the ability of science to solve every last biological problem, the assembled celebrants betrayed a sense of intellectual disarray. Not only was there no unanimity among them about what life was, there was also no shared conviction that the question was even answerable. Some of the participants suggested that it was the wrong question to be asking, while still others floundered around like beached whales in what was apparently a hopeless attempt to come to grips with it.

What their confusion showed, perhaps, was that Schrödinger's Question — "What is life?" — was not in fact a scientific problem, but rather a philosophical one. A scientific question, as everyone knew, was one that could be settled by experiment, or at least by means of a well-defined and objective decision procedure. Whether a substance was acid or alkaline or a number prime or nonprime could be determined by simple tests. But there was no imaginable experiment or algorithm that could reveal what life was; rather, it seemed to be a matter of definition. On the other hand, no attempt to frame a proper definition stood a chance of succeeding if it didn't give due attention to what science had discovered about the nature of life, its processes, and their molecular underpinnings. The question, then, seemed to reside off in some trackless no-man's-land between science and philosophy, and appeared to be at best semidecidable. At any rate, it seemed to pulverize the minds of a few of the scientists who bothered to take it seriously.

For example, when Manfred Eigen, a Nobel Prize winner in chemistry, addressed Schrödinger's Question at the conference, he said: "Not only is this a difficult question; perhaps it is not even the right question. Things we denote as 'living' have too heterogeneous characteristics and capabilities for a common definition to give even an inkling of the variety contained within this term. It is precisely this fullness, variety and complexity that is one of the essential characteristics of life."

That notwithstanding, Eigen immediately proceeded to do what he had just declared was impossible, isolating "three essential characteristics . . . which are found in all living systems yet known," to wit: self-reproduction, mutation, and metabolism.

Still, "it is certainly more sensible to ask," he said, "how does a living system differ from one that is not alive? When and how did this transition take place during the history of our planet or of the universe as a whole?" Those were *experimental* questions, after all, of the type answerable by science. Unfortunately, they were not Schrödinger's Question.

The paleontologist and evolutionary theorist Stephen Jay Gould was, if anything, even more hostile to Schrödinger's project than Eigen had been. After starting out with an obligatory nod to Schrödinger, saying that *"What Is Life?* ranks among the most important books in 20th century biology," Gould proceeded to smash it to bits, saying first that "much of *What Is Life?*'s intellectual foundation—Delbrück's early ideas on reasons for the gene's stability—turn out to be quite wrong." (In 1992, the geneticist James F. Crow had said that "the stability of the gene is now explained by its complementary internal structure and various error-preventing and error-correcting enzymes.") He lambasted Schrödinger's "key claim for an almost self-evident universality in his approach [as being] both logically overextended, and socially conditioned as a product of his age."

Schrödinger was a child of the "unity of science" movement, Gould said, explaining that this was a fad among a group of Vienna School philosophers who in the 1920s preached the doctrine that all sciences shared the same language, laws, and methods, and that no fundamental differences existed between the physical and biological sciences, or for that matter between the natural and social sciences. Instead, all were one. Schrödinger, who had been born, raised, and educated in Vienna, had apparently been indoctrinated with this "reductive unification" business, an intellectual orientation that had warped his views forever afterward. Gould himself was not a partisan of the unity of science movement. "I deplore its emphasis upon standardization in a world of such beautiful diversity," he said, "and I reject the reductionism that underlies its search for general laws of highest abstraction."

Schrödinger's fixation on the chromosomes and the codescript they contained was, according to Gould, myopic. "The answer to 'what is life?' requires attention to more things on earth than are dreamed of in Schrödinger's philosophy."

In Gould's view, a proper answer to Schrödinger's Question had to embrace the fullness of living things, and not just some tiny filaments in their cell nuclei. You had to consider, for example, the hierarchical nature of biological objects: genes, organisms, species. You had to consider mass extinctions, the Cambrian explosion, Darwinian changes in modern populations, the chordate groups of the Burgess shale. In fact, the truly proper focus would not even be restricted to living organisms: it was best to widen the scope of the quest to include the earth itself, the environment, its changes across time, the contingency of natural events, and so forth, all of which played their allotted role in determining what life was. In short, you had to consider . . . *everything*!

THAT PERSPECTIVE WAS substantially echoed in Stu Kauffman's talk. Stuart Kauffman, M.D., had an extraordinary provenance intellectually. He was a recipient of a MacArthur Foundation "genius" award, held simultaneous positions at the University of Pennsylvania and the Santa Fe Institute, and had a list of publications that ranged across virtually all the new-wave scientific disciplines: artificial life, chaos theory, "emergence," chemical evolution, and so on. He was widely regarded as incredibly bright and as an extremely original (perhaps even somewhat *too* original) thinker.

Kauffman began forthrightly enough by asking, regarding Schrödinger's "brilliant" book, "Is the central thesis of the book right?" And he, too, answered in the negative: "He may have been wrong, or at least incomplete." Schrödinger, after all, had concentrated on the aperiodic solid and the code-script inside it, artifacts that everyone now recognized as the DNA molecule and the information it contained.

"Almost all biologists are convinced that such self-replicating molecular structures and a microcode such as DNA are essential to life," Kauffman said. "I confess I am not entirely convinced."

Not convinced?

"The formation of large aperiodic solids carrying a microcode, order from order, may be neither necessary nor sufficient for the emergence and evolution of life," Kauffman said. "In contrast, certain kinds of stable collective dynamics may be both necessary and sufficient for life."

The viewpoint that DNA was neither necessary nor sufficient for the existence, propagation, and evolution of life was not exclusive to Kauffman and was not in fact new.[1] One of the basic theoretical insights of the artificial life movement, going back to at least the mid-1980s, had been the suggestion that DNA was merely a "frozen accident," a chance creation that had persisted through time and across species not because it was the best, or even the only, candidate for the job, but because it had arisen accidentally, had been stabilized by the environment and by the laws of inheritance, and then got grandfathered in to all subsequent organisms that had descended from the first one. DNA's omnipresence among earthly life-forms, in other words, was not a sign that it was somehow inherent to the existence or nature of life, but merely that it was an artifact of life's origin and the common genetic descent from it of all living things. In and of itself, DNA was no more necessary to life than driving on the right-hand side of the road was necessary to driving. (Indeed, part of the motivation behind the search for extraterrestrial life, the artificial life movement, and synthetic biology, particularly the attempt to construct a protocell, was to discover whether life could exist without DNA.)

And so while Schrödinger was right in saying that an aperiodic solid containing a code-script was at the core of life on earth, it did not follow, Kauffman said, that such a thing was an inescapable element of life per se. You didn't need a molecular template—which was what DNA was—in order for life to originate or to evolve.

Kauffman's own view was that life was an emergent phenomenon, a by-product of lots of molecules, enzymes, and catalysts gathering together, shuffling around at random, and then, following the bonding rules that governed the behavior of complex chemical reaction systems, yielding up something entirely new, something that was in the original mixture as only a possibility. As the system increased in complexity and diversity, life would emerge as a chemical phase transition— on the crest of a breaking wave, as it were.

"As the diversity [of the components] increases, a larger number of reactions are catalyzed by molecules in the system," Kauffman said. "At some point as the diversity increases, a connected web of catalyzed reactions springs into existence. The web embraces the catalysts themselves. Catalytic closure is suddenly attained. A 'living' system, self-reproducing at least in its silicon realization, swarms into existence."

The discouraging element here was Kauffman's hedge phrase, "at least in its silicon realization," for this meant that while all these self-organizing miracles happened, they did so mainly inside computers. Actual, test-tube reality was another matter entirely.

The significance of all this, nevertheless, was that within forty years of Watson and Crick, a major biological theorist was pouring cold water over the whole idea that DNA, RNA, or any other such templating molecule was necessary for life. Self-organization, emergence, autocatalytic mechanisms and their brethren could do it all—or at least some of it.

Darwin, likewise, took his hits. "Since Darwin we have come to believe that selection is the sole source of order in biology," Kauffman said. But Kauffman did not believe that, either. This is not to say that he did not believe in evolution at all: "Natural selection is always acting," he assured his audience. But there was more at work in life than selection: in particular, there was self-organization.

"Darwin did not know the power of self-organization," Kauffman said. "Indeed, we hardly glimpse that power ourselves."

The solution was to rethink evolutionary theory, DNA, and self-organization from the ground up, in an effort to see what was and what was not essential to life.

BY THE END of the conference, then, the only claim of Schrödinger's to stand the test of time was how he had reconciled the processes of life and the second law of thermodynamics. But this did not get anyone closer to answering the question What is life?

In 1995, also in commemoration of Schrödinger's book, the biologist Lynn Margulis and her son Dorion Sagan published a *What Is Life?* volume of their own. The book seemed in large part to be an attempt to define life by complete enumeration, as the authors surveyed practically the whole of the living universe, everything from mycoplasma and cyanobacteria (né "blue-green algae") to the planet Earth itself, which the authors viewed as a global, self-maintaining living system ("Life does not exist *on* Earth's surface so much as it *is* Earth's surface," they wrote), and everything in between.

Moreover, many of their chapters ended with the question, "So, what is life?" to which Margulis and Sagan supplied a great and diverse multiplicity of answers, many of them figurative, flowery, or metaphorical—and in some cases memorably orphic. Life, in their view, seemed to be no one thing. Rather it was "planetary exuberance, a solar phenomenon . . . It is matter gone wild . . . a question the universe poses to itself in the form of a human being . . . the representation, the 'presencing' of past chemistries . . . It is the watery, membrane-bound encapsulation of space-time . . . animals at play . . . a marvel of inventions for cooling and warming . . . the transmutation of sunlight," and on and on. As a response to Schrödinger's Question, these were entirely too many answers.

What was clear at this juncture was that despite DNA's success in providing a basis for explaining heredity, the construction of proteins, and therefore the development and growth of organisms, it did not by itself tell us what life was. Plainly, life was something above and beyond, or in addition to, the structure and function of the gene. But what?

In 1985, also in the context of reconsidering Schrödinger's Question, the mathematical physicist and origin-of-life theorist Freeman Dyson made note of the extremely slight attention Schrödinger had paid to another important life function, metabolism. Dyson claimed, in fact, that there was a logical gap between Schrödinger's discussion of the gene and its code-script on the one hand and the metabolic functions of organisms on the other.

"Looking back on his 1943 lectures now with the benefit of forty years of hindsight, we may wonder why he did not ask some fundamental questions which the gap might have suggested to him. Is life one thing or two things? Is there a logical connection between metabolism and replication? Can we imagine metabolic life without replication, or replicative life without metabolism?"

Dyson charged that DNA's undeniable proficiency at explaining replication had pushed the whole concept of metabolism into the shadows. Life had become virtually synonymous with the replication of cells and the reproduction of whole organisms. But there was more to life than those two functions. For one thing, not all organisms reproduced: sterile hybrids such as mules did not but were alive nevertheless. For that matter, many nonsterile organisms, whether by accident or design, never reproduced themselves, either, but were no less alive on that account. Moreover, most of the body's individual organs—the heart and the brain, for example—were "alive" in some derivative sense, despite the fact that they had no capacity whatsoever for reproduction. And even some of the body's cells, brain cells in particular, did not reproduce, but were nonetheless living entities. But all of those nonreproducing things—organisms, organs, and cells—were kept alive by metabolic processes.

Most important of all was the fact that metabolism seemed to be an even more basic life function than replication, development, or growth were. None of those latter activities could take place if the organism in question lacked a working metabolism. They were parasitic on metabolism; metabolism was not parasitic on them.

"This logical analysis of the functions of life," Dyson said, "helps to explain and to correct the bias toward replication which is evident in Schrödinger's thinking and in the whole history of molecular biology . . . In the balance of nature there must be the opposite bias."

Perhaps it was metabolism, then, that constituted the core and basic nature of life, and was in fact the answer to Schrödinger's Question. Life was not replication, not reproduction, not a code-script, not the gene. Most fundamentally, life was metabolism.

Five

ATP and the Meaning of Life

METABOLISM IS what makes organisms go. It converts food into bodily structure and kinetic energy, two things that living entities cannot do without. Anyone who felt so inclined could forgo reproduction for their entire lives, but no one could do without metabolism for a moment without extremely serious consequences. Between the two of them, then, which seemed to be closer to the sum and substance of life?

Whether on the cellular level or that of the whole animal, the workings of metabolism were terra incognita to most people. The average person's understanding of the human metabolic process was that you took in food and water, excreted waste products, burned up energy, and perspired. The greater part of all this frantic activity took place somewhere down in the black holes of the intestinal tract, where the biologically useful substances, whatever they were, entered the blood-stream and somehow produced energy.

That was about as far as the layman's metabolic notions went. Despite this knowledge deficit, however, people often tried to affect their metabolism, "boosting" it or "speeding it up" by drugs, specific food types, herbs, dieting, or exercise in an attempt to control their weight. They tried to "burn fat fast" so as to "drop twenty pounds in only four weeks," and the like, as promised by countless ads.

In general, metabolism was the process by which an organism maintained itself as a living entity. It did this by bringing in raw materials from outside itself, reassorting their chemical constituents, breaking down certain molecular structures and building up others from their component molecules, thereby converting the foodstuffs into substances required for the organism's continued existence, as well as for its energy supplies. Your metabolism was responsible for the fact that you could eat breakfast, wait awhile for it to "digest," and then go out and jog five miles.

That metabolism could in fact be the biological essence of life was plausible in the sense that the body's motion and activity, its incessant molecular reshuffling, all of its chemical getting and spending, depended upon the conversion of nutrients into energy. When metabolism stopped, death ensued, complete with rigor mortis.

Metabolism, furthermore, appeared to perform on a daily basis, and automatically, the extraordinary feat of transforming the nonliving into the living. By and large, people ate dead things (such as cooked plants, fish, fowl, and meat), which their metabolic systems then converted into integral and indispensable parts of the living organism. Somewhere in the middle—somehow—chemistry became biology.

Metabolism also explained how it was that living things appeared to violate the second law of thermodynamics, but didn't. As Schrödinger argued, their metabolic systems merely withdrew orderly energy out of the environment and put disorder, or entropy, back into it.

But the story of how an organism's metabolism managed to do all these things at once was a tale of some complexity.

METABOLISM HAD an original pioneer and two main heroes. The pioneer was a Venetian physician by the name of Santorio Santorio (1561–1636). Santorio was the eldest son of Antonio and Elisabetta Santorio, wealthy nobles who lived in a time and place where it was fashionable to bestow the family name as the given name upon one's firstborn son. Santorio lived an unusual life and at the end of it had made important, if in some cases rather strange, contributions to medicine as a science.

He lived in an era when the theory of bodily humors was still in vogue, a doctrine that went back to Hippocrates and postulated that health consisted of maintaining the proper balance of the four humors: blood, phlegm, choler, and black bile. This was an unscientific doctrine for two main reasons. First, it rested on no empirical evidence whatsoever; second, the influences which these substances (or alleged substances) had upon health could not be measured by any experimental test. Members of the medical profession in those times also held several quaint physiological notions that seemed only "obvious" and "natural," for example, that people who lived in tropical climates had higher average body temperatures than those who lived in more northerly latitudes. But since no thermometers existed, that belief could not be tested either.

It was Santorio's mission in life to introduce some objectivity and experimentalism into the rather vapid field of medicine as it then stood. He was friendly with Galileo, who had used a device called a thermoscope to record changes in air temperature. The thermoscope lacked a scale, however, which meant that temperature observations could be made only comparatively, or qualitatively, not absolutely, or quantitatively. Santorio added to the thermoscope a graduated scale divided into equal units between the temperature of snow and that of a candle flame, transforming the device into a thermometer, an instrument that could be used to make an objective measurement of human body temperature. He also invented a "pulsiloge" (pulse timer) with which to measure a patient's pulse rate. These and other instruments were some of the first devices that took observations of internal bodily states and systematically correlated them with publicly observable and measurable conditions in the external world.

Still, Santorio's greatest achievement was yet to come. This was an attempt to observe, measure, and record some of his own metabolic processes over a substantial period of time. Health, he thought, was a matter of establishing an appropriate symmetry between what went into and what came out of the body (a notion that, in some form, persists even today). He decided to measure those factors by weighing himself on a scale—a traditional balance-beam mechanism. The idea of putting people on scales was revolutionary for its time. In those days, scales were not used to weigh human beings, but were reserved for objects of commerce such as meat, poultry, coins, herbs, candles, gems, gold, and the like. Santorio nevertheless fashioned a device by which he could weigh himself, his food, and his excreta, and monitor these variables not just for a day or two, but for years on end. He worked, slept, had sex, in effect practically lived on his balance-beam contraption. As one science historian put it, Santorio "minutely recorded his own variations in weight for over thirty years, having devised domestic static tools—including a static bed—and transformed part of his living space into a scale."

Santorio was the first systematic weight-watcher in history, and even invented a clever if relatively coercive method for staying on a "diet." You would place on the opposite end of the scale a counterweight corresponding to your own weight plus that of the quantity of food you wanted to consume. Then, while seated in the weighing chair, you would begin to eat, taking the food and drink from a table that was separate from the balance-beam mechanism. At the precise moment at which you had ingested your predetermined allotment of food, the weighing chair would drop down past the table and forcibly end your meal.

After an incredible three decades aloft upon his personal weighing spaces in Venice and Padua, Santorio produced a treatise, *Ars de statica medicina* (1614), in which he stated his conclusions, among which were that the weight of his excreta was a great deal less than the weight of his food. The rest wafted away into the air as "insensible perspiration," he said, by which he meant water vapor eliminated through the skin pores or through the nose and mouth, the latter of which could be observed by breathing onto a mirror.

His book, which was written in the form of some five hundred aphorisms ("Aphorism IV: Insensible Perspiration alone, discharges more than all the servile Evacuations together"), became a sensational bestseller, was reprinted forty times and translated from the Latin into Italian, French, German, and English. It was arguably the first diet-craze book in history, and was an especially big hit in England, where it met with a variety of reactions, including parody and ridicule, but also attracted its share of followers. One Englishman, who wanted to keep his weight at two hundred pounds, said he had not left his Santorian weighing chair in three years. By the end of it he confessed that in pursuing his health "by Ounces and Scruples," he had removed himself from polite society, and dined not "by the Clock," but only by the chair.

Primitive though they might have been, Santorio's chronicles of his bodily weight changes, and of his influxes and effluxes, constituted the first quantitative record of metabolic activity in the annals of medicine.

IN MODERN TIMES, the study of metabolism became a specialty province within biochemistry and constituted a discipline with a complexity that rivaled any other in science. Not only was the overall process difficult to grasp, but some particular metabolic subprocesses were themselves so complicated, tortuous, and involved so many sequential chemical reactions that they made lesser life processes look simple by comparison. Glycolysis ("sugar breakdown"), for example—the conversion of a simple sugar into its by-products—involved ten separate steps, each mediated by a different enzyme, a collective zoo of biomolecules with forbidding, perhaps even frightening, names such as phosphofructokinase, phosphoglyceromutase, and glyceraldehyde phosphate dehydrogenase, among others. Notwithstanding all of this dark mumbo jumbo, one biochemist described the convoluted twists and turns of glycolysis as "ultimately quite elegant." Plainly, the study of metabolism was not for those with weak nerves.[1]

Still, some basic metabolic terms and distinctions were simple enough. Total metabolism included all the biochemical processes of an organism, and these were of two basic types: the breaking down of complex organic molecules into smaller units (catabolism), and the synthesis or building up of such structures from smaller components (anabolism). The growth of an organism occurred when anabolism exceeded catabolism; weight loss resulted when the opposite was true. When the two processes were in balance, tissue mass remained the same.

Energy metabolism was concerned with the net heat production of an organism, while "intermediary metabolism" dealt with the chains of chemical reactions that took place within cells. Those reactions progressed along various discrete metabolic "pathways," which were sequences of enzymatic steps that transformed foodstuffs into bodily structures and/or energy.

The overall metabolic activities of most higher organisms occurred in stages, the first of which was the lowly and familiar "digestion." This was the initial gross breakdown of foods, which were primarily a mix of proteins, fats, sugars, and starches (also known as polysaccharides). In human beings, digestion took place mainly in the stomach and intestines, and the process was helped along by gut bacteria that secreted enzymes, substances that transformed the proteins into their twenty different amino acids, starches into sugars, and so on.

After those first baby steps, matters got slightly more complicated. Somewhere downstream from digestion, its by-product molecules entered a gigantic Ferris wheel of molecular reactions known as the Krebs cycle. The Krebs cycle was named after Hans Adolf Krebs, the German biochemist who worked out the series of its steps in 1937.

Hans Adolf Krebs, born in Hildesheim, Germany, in 1900, was the son of an otolaryngologist, and he initially wanted to practice medicine himself. But even before he got his M.D. degree in 1923, he had published a scientific paper on techniques for staining muscle tissue, and had decided on a career in pure research. As he worked his way through various labs, including that of Otto Warburg at the Kaiser-Wilhelm Institut in Berlin, he became increasingly interested in studying intermediary metabolism, the metabolism of cells.

"I felt greatly attracted by the problem of the intermediary pathway of oxidations," he explained much later. "These reactions represent the main energy source in higher organisms, and in view of the importance of energy production to living organisms (whose activities all depend on a continuous supply of energy) the problem seemed well worthwhile studying."

Comparatively little was known at the time about the series of individual biochemical reactions that started when foodstuffs entered the body and ended with the production of new tissue structures and energy. It was Krebs's goal to establish the unbroken sequence of metabolic reactions that occurred in cells.

Meanwhile, in Szeged, Hungary, Albert Szent-Györgyi, who in 1932 had discovered the chemical makeup of vitamin C, was carrying out some of the preparatory work by studying cell respiration, the process by which cells took up and utilized oxygen, an important part of the metabolic process in aerobic organisms. In this effort he used a novel experimental medium: minced pigeon breast muscles. Since these were the bird's chief flight muscles, they were exceptionally powerful for their weight, and therefore they made ideal test specimens for studying oxygen take-up.

Through a series of experiments, Szent-Györgyi unearthed part of a critical metabolic pathway, the chemical transformation of lactic acid to carbon dioxide. He found that any one of four acids (malic acid, succinic acid, fumaric acid, and oxaloacetic acid) could act as catalysts in that reaction. Further, Szent-Györgyi postulated that the overall transformation was cyclical, meaning that it continued on as new inputs were changed into new outputs. This, then, was the start of a small Ferris wheel.

Krebs, who knew Szent-Györgyi personally, began his own experiments with minced pigeon breast muscle and discovered additional steps in the overall enzymatic reaction pattern that prompted the full chain of chemical conversions of one substance into the next. He found, among other things, that citric acid played an essential role in the process, and for this reason the complete series of steps became known as the citric acid cycle. But because his own contribution to understanding these reactions was so great, by 1941 it was already being called the Krebs cycle, by which name it is generally known today.

Fundamentally, the Krebs cycle's series of molecular conversions was responsible for breaking down carbohydrates, fats, and proteins into carbon dioxide and water in order to generate energy. The full procession of events was a show

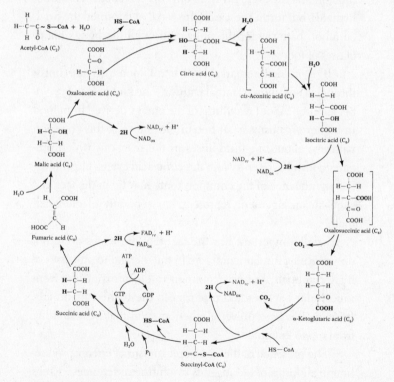

The Krebs cycle

piece of chemical complication and is best appreciated in animated versions, as can be found on some websites (where it is sometimes known as the tricarboxylic acid cycle).[2]

Even Krebs himself wondered why the cycle was so circuitous and complex, but in the end he decided that, as complicated as it was, the citric acid cycle was probably the most chemically efficient method for extracting the available energy from a wide variety of nutrients. Krebs had initially thought that the cycle involved only the metabolism of carbohydrates, but further research showed that it formed the final, common pathway for the utilization of all the body's major classes of foods.

"It is indeed remarkable that all foodstuffs are burnt through a common terminal pathway," he said in his 1953 Nobel lecture. "About two-thirds of the energy derived from food in higher organisms is set free in the course of this common pathway; about one-third arises in the reactions which prepare foodstuffs for entry into the citric acid cycle. The biological significance of the common route may lie in the fact that such an arrangement represents an economy of chemical tools."

Equally important was the fact that the Krebs cycle not only played a fundamental role in the metabolic processes of higher animals, but was common to *all* forms of life, from unicellular bacteria and protozoa to the highest mammals. It seemed to be a universal phenomenon that pervaded life from top to bottom.

"The presence of the same mechanism of energy production in all forms of life suggests two further inferences," Krebs said. "Firstly, that the mechanism of energy production has arisen very early in the evolutionary process, and secondly, that life, in its present forms, has arisen only once."

But the true payoff of the Krebs cycle for any living organism lay at the far edge of a tiny epicycle that whirred inside the main cycle, generating a small fountain of molecules identified by the letters ATP. In fact, each turn of the Krebs cycle generated as many as thirty-eight molecules of that critical substance.

IF HANS ADOLF KREBS WAS the first hero of metabolism, the ATP molecule was the second. ATP, adenosine triphosphate, was one of the most important molecules found in living things because it was the storehouse of cellular energy: it propelled the contraction of muscle cells, it powered the transmission of nerve impulses, it was used in the construction of proteins and in the chemical breakdown of foods. If there were truly a "spark of life" molecule, one that animated living matter and was the ultimate driving force of all life, it would be adenosine triphosphate. As the chemist P. W. Atkins has said, "Where there is life, there is ATP."

Adenosine triphosphate was discovered as a distinct substance in 1929 by three scientists working independently, but its chemical structure was not deciphered until 1935. The molecule consists of a two-part unit comprising adenine (a ring of carbon and nitrogen atoms) and a ribose (a sugar), which when bonded together form a compound known as adenosine. That unit is then connected to three identical phosphate groups in succession (each of which is itself composed of a phosphorus atom and four atoms of oxygen, along

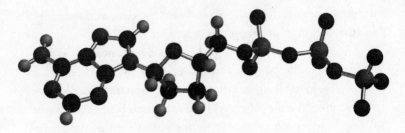

ATP (adenosine triphosphate) molecule

with their associated hydrogen atoms), all of the individual units being linked together in a chain.

Starting from the ribose, the phosphate groups are referred to as the alpha, beta, and gamma phosphates. That chain of phosphates is extremely rich in chemical energy, particularly in the high-energy covalent bond between the terminal phosphate and the rest of the molecule—which is to say, between the beta and gamma phosphate groups, in a so-called pyrophosphate bond. When the terminal phosphate separates from the rest of the molecule at the demand of an enzyme, a pulse of energy is released. In essence, this is the energy that powers all of life and its processes.

With its gamma phosphate removed, the original ATP molecule is converted to adenosine diphosphate, ADP, a dangling phosphate (P), and energy:

$$ATP \rightarrow ADP + P + energy$$

Instructors in biology courses often draw an analogy between ATP molecules and rechargeable batteries. As the battery's energy is used, its energy state is lowered until it reaches the point where it cannot be used again until it's recharged. ATP is the charged battery, ADP is the discharged version, while the input of additional energy (in the form of a new phosphate bond) "recharges" ADP into ATP once again, in a process known as oxidative phosphorylation.

Any given ATP molecule is recycled—converted to ADP and back again to ATP—some 2,000 to 3,000 times a day. ATP synthesis takes place in the mitochondria, tiny sausage-shaped organelles located in the cell's cytoplasm. The mito-chondria are such tiny structures that it is said a billion of them clumped together would be no bigger than a grain of sand.

The entire process of ATP synthesis is so important to life's functioning that any significant interference with it, as, for example, in cases of potassium cyanide poisoning (which blocks cellular uptake of oxygen), causes death within min-utes. ATP is what gives the muscles their flexibility, and when death occurs, the resulting rigor mortis is due to a deficiency of new ATP within the cells. (Rigor sets in more quickly if the body's previous supplies of ATP have been depleted, as dur-ing exercise or in a fight.) Fertilizers contain large amounts of phosphate in order to provide for the ATP synthesis that oc-curs even in plant cells.

With the ATP molecule we have reached a bedrock point to metabolism, and have perhaps identified one of the under-lying principles common to all of life. Life's motive force, its powerhouse, the very engine that makes life go, has been reduced to the ordinary laws of physics and chemistry—specifically, to a known molecule and the repeated bonding and separation of its constituent atomic groups. From this result it might appear that not many questions remained about the existence and nature of life. But that appearance would be deceptive.

For one thing, the fact that ATP and the Krebs cycle were universal among the natural life-forms of earth did not mean that either or both of these phenomena were necessary conditions of life per se. Just as it was possible that DNA was a "frozen accident" and not an inherent component of life itself, so too might the metabolic processes and products of biological life on earth be merely contingent artifacts of history. Extraterrestrial life, if it existed, might well be based upon genetic molecules other than DNA, and metabolic machinery other than ATP and the Krebs cycle. (Likewise, it remained an open question whether carbon-based life is the only type of life there is.)

Indeed, if the protocell project was successful, it would demonstrate that life was broader in nature, that it was a wider, more general type of phenomenon than was common to the life-forms of earth. With respect to earthly life, however, as parochial to its time and place as it might be, two questions yet remained: How did it begin? And, once begun, how did it grow and diversify into the variety of life-forms that surround us?

Six

Origins

THERE IS probably no more disputed problem in theoretical biology than the question of how life arose on earth. The field is crowded with fads and fashions, "standard models," flavor-of-the-week theories, countless "scenarios" as to how it might have happened, pie-in-the-sky ideas and wild speculations, and some outright flakiness—all of it the product of competent scientists, some of whom are even Nobel Prize winners. Indeed, the dispute over life's origins might never be settled definitively, for the earliest organisms, whatever they were, left no physical traces of their existence, and so the evidence of their genesis has long since departed the scene. As Francis Crick once said: "I cannot myself see just how we shall ever decide how life originated." Steen Rasmussen went even further, saying that when it comes to the question of how life originally began on earth, "nobody has a freakin' clue."

Within the Western scientific tradition, the initial "wrong" theory of life's origin was spontaneous generation, the notion that primitive life-forms arose out of various types of decomposing matter. This doctrine was a hotbed of controversy for several centuries. False as it turned out to be in its original incarnation, the theory made a certain amount of horse sense. For one thing, there seemed to be a fair amount of observational evidence in favor of it: with their own eyes, people saw mice, flies, spiders, worms, and the like crawling out of mud, decaying straw, rotting grain, or other offal, and decided that those life-forms had by some unknown process spontaneously arisen in those founts of bad smells and noxious effusions. The best testimony was the case of maggots, which seemed to materialize directly from putrefying meat (when in fact flies had deposited their eggs there).

Second, when scientists tried to disprove the theory, they were quite unsuccessful—at least at first. In 1748, for example, two naturalists named John Needham and Georges Buffon boiled a quantity of mutton broth and then sealed it up in glass containers. Supposedly, the heat would kill any microorganisms that had been present in the broth initially. But when the glass containers were opened a few days later, rich colonies of microorganisms were observed through the microscope. Where or what could they have come from but nonliving matter?

Twenty years after that, the Italian biologist Lazzaro Spallanzani was *apparently* successful in refuting the theory when he improved upon these experiments by boiling the broth for longer periods—for as long as forty-five minutes or so, killing not only the organisms but also their spores, eggs, seeds, or whatever else, and not only those in the broth but also those in the air inside the flask and on the inner surfaces of the container. No organisms ever cropped up in the sealed containers after such prolonged boiling, and from this it should have followed once and for all that the doctrine of spontaneous generation was false.

But supporters of the theory had an irresistible comeback, saying that the longer boiling period had killed not only the microorganisms and their spores, but something even more significant: an impalpable "vital principle," "life force," or *vis viva* that resided in the very air itself and was the essence and original wellspring of life. The great virtue of this theory was that it seemed immune to criticism because there looked to be no conceivable way to test it.

But there was. In 1860, Louis Pasteur repeated the boiling experiments using what came to be called his "swan-necked" flasks. These were ordinary spherical lab flasks whose necks had been narrowed down, drawn out to extreme lengths, and then bent through various sharp curves and angles. The necks were open to the air at the end, but the aperture was very small, only one to two millimeters in diameter, about the width of a plastic cocktail straw. The length of the necks, plus the nature and degree of their curvature, made it difficult for airborne spore-carrying dust motes to enter the necks and then migrate up into the boiled liquid.

In his most elegant experiment, Pasteur placed yeast infusions into a group of such flasks, boiled some of them and left several others unheated as controls, then let them all stand exposed to the open air for two to three days. At the end of that period, the unheated yeast infusions were covered with mold while the boiled ones showed no traces of any form of life. Pasteur's conclusion was that the "sinuosities and inclinations" of his swan-necked flasks prevented spores and bacteria-carrying dust motes from reaching the liquid. Still, the flasks were open to the air and so any postulated "vital principle," "vegetative force," or other occult influences should have been able to wend their way through the open neck and into the liquid, where they could in their magical fashion restart life.

Spontaneous generation was pretty much a dead letter after that, as Pasteur himself announced at a gala lecture he delivered at the Sorbonne on the evening of April 7, 1864. In attendance were Alexandre Dumas, George Sand, and Princess Matilde Bonaparte, among other luminaries, all of whom had come to hear the great scientist speak on the subject.

"Can organisms come into the world without parents, without ancestors?" Pasteur asked his audience. "That's the question to be resolved."

He rehashed the history of the debate and described some of the experiments that had been performed by others before him. Then he recounted his own series of tests, including those with the swan-necked flasks. There seemed to be no question as to their outcome or ultimate significance.

"I have removed life, for life is the germ and the germ is life," Pasteur said. "Never will the doctrine of spontaneous generation recover from the fatal blow that this simple experiment delivers to it." To believe otherwise, he said, was to believe that matter can somehow organize itself into a living being. "Can matter organize itself?" he asked. Pasteur's answer was no.

By the middle of the twentieth century, though, the answer was once again yes.

SOON AFTER Pasteur's bravura demonstrations, the German biologist August Weismann asserted the "continuity of the germ plasm" doctrine, which stated that life always came from life and that there was no spontaneous creation of any organism. Life did not arise from nonlife, except perhaps at the very beginning. As to how that might have happened, way back when, two eminent scientists of the period put forward essentially the same suggestion. The first was Pasteur himself, who, a good Catholic and a believer in a divine creation, brought up the theory only to reject it. He laid out the scenario he opposed, saying: "Take a drop of sea water . . . that contains some nitrogenous material, some sea mucus, some 'fertile jelly,' as it is called, and in the midst of this inanimate matter, the first beings of creation take birth spontaneously." But to Pasteur, life did not just pop into existence; God created it.

The second scientist was Charles Darwin. Officially and for the public record, Darwin stated in the *Origin of Species* that "life, with its several powers, [had been] breathed by the Creator into a few forms or into one." Privately, however, in a letter written in 1871, he spoke of life having possibly arisen "in some warm little pond, with all sorts of ammonia and phosphoric salts, lights, heat, electricity, &c. present." Perhaps under those conditions, he said, "a protein compound was chemically formed, ready to undergo still more complex changes." Darwin meant it seriously.

By the mid-twentieth century, the established scientific doctrine regarding the origin of life on earth was that it had indeed spontaneously arisen by some sort of chemical self-assembly process that took place in a "prebiotic soup." That theory became popular in the wake of Stanley Miller's well-known 1953 closed-flask test. Performed as part of his requirements for a doctoral degree, this crucial experiment more or less replicated the conditions described by Darwin. Working in Harold Urey's lab at the University of Chicago, Miller placed less than a pint of water inside a small glass flask that was part of a closed recirculating loop that also included a separate five-liter flask into which he added ammonia, hydrogen, and methane—three gases that he and Urey believed were present in the atmosphere of the early earth. Miller boiled the water to send steam and the other gases through the loop, while at the same time discharging electrical sparks through the five-liter flask in order to simulate lightning.

At the end of a week's worth of this, he found that the system contained twenty chemical compounds that were not there originally, and that four of them, glycine, alanine, glutamic acid, and aspartic acid, were amino acids—the building blocks of proteins, which in turn were the building blocks of living things. After the manner of Pasteur, Spallanzani, and the others before them, Miller said in his 1953 *Science* report on the experiment that "the amino acids are not due to living organisms because their growth would be prevented by the boiling water during the run."

Other scientists later criticized Miller's experiment on the ground that the prebiotic atmosphere was not in fact what he and Urey thought it was. The essential point, however, remained unchanged—namely, that a collection of inert chemicals could under the right conditions yield some of the basic protomolecules of life.

Still, it was a long way from the production of a few random amino acids out of a carefully designed, primed, and prepared chemical soup to the creation of an actual independent living organism. Indeed, amino acids were fairly easy to create, as the Urey/Miller experiment amply demonstrated, but there was more to life than amino acids.[1] Whether life was metabolism, replication, evolution, or all three, the fact was that all of those phenomena required the existence of far more complicated substances than amino acids, substances such as structural proteins (needed for cell construction), catalytic proteins or enzymes (needed for replication and for metabolism), and some sort of template molecule, such as DNA or RNA (needed for information storage, replication, and protein synthesis). An elementary exercise in numerical combinatorics sufficed to show the scope and dimension of getting the right proteins out of the amino acid building blocks. A protein consisted of many separate amino acids bound together in a chain. If a hypothetical protein was 200 amino acids long (which was not exceptionally long for a protein), then, given the fact that there were 20 different amino acids that could occupy each of those 200 spaces, there were 20^{200} possible amino acid combinations, which was a number approximately equal to 10^{260}. By any standard, that was a big number; the number of elementary particles in the universe, by contrast, was thought to be "only" 10^{80}.

If there were so many possible combinations, then what were the chances that the original molecules of life had arisen spontaneously, by means of amino acids randomly bumping into each other in an ancient primeval sea? According to some scientists the answer was, practically nil. Francis Crick, for example, held that the "organized complexity" of life "cannot have arisen by pure chance," and that "life, from this point of view, is an infinitely rare event."

On the other hand, why was 10^{260} a relevant number? All it described was the total number of possible amino acid combinations in a protein that was 200 amino acids long. What if there was some sort of chemical selection pressure operating that preferentially caused some subset of those possibilities to be favored over others? Also, nothing in the nature of life required that all those 10^{260} possibilities be actualized. What life required in order to exist was some finite number of proteins, which meant that it needed *some* combinations of amino acids to work, not all possible ones. The fact that the earliest organisms appeared some 3.85 billion years ago, whereas the earth itself didn't even become solid until 3.9 billion years ago, meant that life arose very early, practically as soon as it possibly could, which in turn meant that it couldn't have been all that hard for life to organize itself into existence.

What, then, propelled it to do so? Stu Kauffman's answer was a process called autocatalysis.

Kauffman was of the view that life, far from being rare or

unlikely, was, as it were, dying to get born at the first available opportunity. Chemically, there was a perfect method of accelerating the process of molecular evolution, and that was by means of catalysts. Catalysts sped up chemical reactions in a way that offered possible shortcuts to the genesis of life. In a series of books and articles that went back to the mid-1980s, Kauffman suggested that the original prebiotic soup may have contained an *autocatalytic network*, a collection of diverse chemical molecules and catalysts in which some molecules catalyzed the creation of others until, when a certain critical point was reached, the whole collection of them was catalyzing itself. For example, a batch of molecules of Type A catalyzed the formation of others of Type B, which in turn catalyzed still others of Type C, and so on around and around in a big whirring loop at the end of which molecules of Type Z were catalyzing those of Type A—all of this happening continuously and simultaneously. In this way, something analogous to a protometabolic system, a primitive form of life, bootstraps itself into existence. (In gross mechanical terms, this was like starting your car by using the electrical energy stored in the battery. The engine turns over and ignites, after which the alternator sends electrical energy to the battery again in a self-starting, self-sustaining loop.)

"The secret of life, the wellspring of reproduction, is not to be found in the beauty of Watson–Crick pairing," Kauffman announced, "but in the achievement of catalytic closure . . . If a sufficiently diverse mix of molecules accumulates somewhere, the chances that an autocatalytic system—a self-maintaining and self-reproducing metabolism—will spring forth becomes a near certainty."

Well, maybe: there always seemed to be a bit of hocus-pocus in this. Other scientists, including Francis Crick himself, took a relatively dim view of such "bursting at the seams" views of life's origins. "It is impossible for us to decide whether the origin of life here was a very rare event or one almost certain to have occurred," Crick said in his 1981 book, *Life Itself: Its Origin and Nature.* "Even though arguments are sometimes put forward for the latter view, they seem very hollow to me."

On the other hand, Crick's own view was hardly more confidence-inspiring: he argued that life on earth had originated in outer space and was brought here on a rocket ship.[2] Crick was not joking about this, and in fact the theory of extraterrestrial origins was not new with him but had been advanced in 1908 by the Swedish chemist Svante Arrhenius, who saw living spores wafting across the universe propelled by beams of starlight in a process called panspermia. Crick (together with his colleague Leslie Orgel) merely added the rocket ship and renamed the process *directed* panspermia. That a Nobel Prize–winning scientist of Crick's standing would embrace what might be called a desperation theory of life's origins only reflected what he saw as the seriousness of the obstacles that stood in the way of accounting for the rise of life by more conventional means: the problem of getting the right proteins from the almost endless possible combination of amino acids; the problem of explaining the origin of DNA, RNA, and the genetic code; and the further problem of getting each of these individually unlikely things to happen together and join up at the right place and time.

"An honest man, armed with all the knowledge available

to us now, could only state that in some sense, the origin of life appears at the moment to be almost a miracle, so many are the conditions which would have had to have been satisfied to get it going," Crick said. "The plain fact is that the time available was too long, the many microenvironments on the earth's surface too diverse, the various chemical possibilities too numerous and our own knowledge and imagination too feeble to allow us to be able to unravel exactly how it might or might not have happened such a long time ago, especially as we have no experimental evidence from that era to check our ideas against."

The question was whether his Flying Saucer Theory of life's origin, to which he devoted an entire chapter of his book, was a reasonable alternative. His was at best a theory of life's arrival on the planet rather than a theory of how life originally arose, wherever, whenever, and by whatever means that might have happened.

Still, absolutely the last word on the subject of "life from outer space" came from Lynn Margulis and Dorion Sagan, who in their 1995 book *What Is Life?* quipped that "Earth itself is suspended in space, so any way we look at it, life came from space."

TAKEN TOGETHER, there was an overabundance of theories as to how life arose on earth ("too much speculation running after too few facts," Crick said). There was the iron-sulfur world theory; the primordial sandwich theory; the theory that life arose in clay; in deep-sea vents; in oily bubbles at the seashore; et cetera. But any theory of life's origin had to face a severe technical hurdle in the form of a chicken-and-egg dilemma. Modern cells were based largely on two species of biochemical molecules: DNA and proteins. But DNA could not replicate without proteins (enzymes made out of proteins started the unzipping of the double-stranded DNA molecule), whereas proteins themselves could not exist without DNA molecules (which contained the recipes for making the proteins). It was a classic Catch-22, expressed on the level of the original molecular constituents of life. Each kind of molecule depended on the other one for either its existence or its functioning, and so it seemed as if the processes that were at the very core of life could not possibly have gotten off the ground.

In the late 1960s, three independent theorists (Francis Crick, Leslie Orgel, and Carl Woese) suggested a possible way out of the dilemma. They hypothesized the existence of a dual-purpose type of molecule, one that could exist in at least two versions and therefore could perform both of the necessary functions: it could act as an enzyme (and thus escape the need for the proteins that acted as catalysts), and it could replicate without the existence of DNA. The molecule that could do all this, they said, was RNA, which was a single-stranded version of DNA, with uracil in place of DNA's thymine. Because it was single-stranded, it could fold up the way proteins did and perhaps even act as an enzyme. And because it was an information-encoding molecule, it could direct life processes as DNA did. Moreover, RNA molecules were capable of being replicated.

The main problem with this idea was that at the time it was proposed, all known enzymes were proteins. But some fif-

teen years later, in 1983, two experimentalists working independently, Thomas Cech and Sidney Altman, found an exception to the enzymes-are-proteins rule: they discovered
RNA molecules that promoted chemical reactions, and
therefore acted as enzymes themselves. Since RNA stood for
ribonucleic acid, Cech and Altman called these new molecules "ribozymes."

Ribozymes made possible a whole new world, one in
which RNA was the predominant biological molecule. In
1986 the Harvard biochemist Walter Gilbert used the term
"RNA world" to describe an early stage in life's genesis in
which "RNA molecules and cofactors [were] a sufficient set
of enzymes to carry out all the chemical reactions necessary
for the first cellular structures." RNA, in other words, could
both replicate itself as well as catalyze the other reactions
needed for life. According to this view, the entire primeval
world of living things was composed of, and made possible
by, RNA.

The RNA-world hypothesis received a major boost in
2001, when two researchers at Yale, Thomas Steitz and Peter
Moore, successfully characterized the major structure of the
ribosome, the cellular component that in modern cells was
responsible for protein synthesis. Using X-ray crystallography,
they showed that the ribosome itself was composed of RNA.
"We have established that the ribosome is a ribozyme, an enzyme in which catalysis was done by RNA, not protein,"
Steitz said. "This means that in the very early days of evolution, protein synthesis evolved using RNA molecules because
there were no protein molecules."

The chicken-and-egg problem, then, was apparently

solved: RNA came first; RNA synthesized proteins; and DNA somehow came along later. Still, however promising all this might have sounded as a model for life's origins, there were a few problems even with the RNA-world hypothesis. For one thing, RNA molecules were fragile and easily broke down into their constituent elements. Their tendency toward chemical disintegration would have been even more pronounced in the harsh environment of the early earth, a circumstance that would sabotage the whole process before it even got started. The second problem was the question of how the RNA world and its component molecules of RNA themselves ever came into existence. Some process of chemical evolution could presumably account for it, but exactly what the mechanism was remained unknown.

And finally there was the matter of metabolism to be considered. The RNA world was essentially a *genetic* apparatus: some RNA molecules functioned as genes, others as catalysts, and still others as ribosomes—the molecular machines responsible for building cellular structures. But if the RNA world, supposing it ever existed, might explain the genetic side of life, it nevertheless left cellular metabolism largely out of the picture. In fact, it also left *cells* out of the picture: how did *they* arise? That is to say, how did the two phenomena, metabolism and the genetic apparatus, ever get combined and smoothly integrated with each other in such a way that led to a metabolizing and replicating independent cell?[3]

Freeman Dyson, the Institute for Advanced Study theorist, maintained a "theory of double origins" that, at least on the surface, seemed to account for both phenomena. According to Dyson, life was not one thing but two, metabolism and replication, and instead of life arising all at once as a whole, each component part arose separately. Dyson advanced this theory in his book *Origins of Life*, published by Cambridge University Press in 1999. While the book went into the double-origin theory in considerable technical detail, it was possible to state the essentials of the theory quite simply.

"I like to call my theory the garbage-bag theory of the origin of life," Dyson said in a recent email. "Life began with little bags of garbage, random assortments of molecules doing some crude kind of metabolism. That is stage one. The garbage bags grow and occasionally split in two, and the ones that grow and split fastest win. Then one of the garbage bags gets a head start by inventing ATP, the molecule which is a good catalyst for almost all reactions. ATP is first cousin to adenosine nucleoside, which is one of the components of RNA. So adenosine nucleoside accumulates in some of the garbage bags and makes RNA. Now RNA replicates and becomes an independent parasitic form of life inside the garbage bags. This is stage two, and it is also the RNA world. The RNA parasites evolve all kinds of RNA structures up to and including the ribosome. Then the ribosomes learn how to make proteins following instructions coded in messenger RNA and take over the metabolism of the cell. The garbage is got rid of. This is stage three, modern life with RNA and proteins running the show together.

"I think this picture, including the RNA world as an intermediate parasitic stage in the history of life, is consistent with all the evidence," he added. "But of course the first stage, the garbage-bag stage, has disappeared and left no evidence that it ever existed.

"That's all I know," he said. "Now you have to make the rest up yourself."

Seven

The Spandrels of San Marco

UNLIKE THE QUESTION of life's origins, which was a scientific free-for-all, the problem of how life evolved, once it arose, had a single, relatively well-defined, largely understood, and factually well-established solution. Indeed, the theory of evolution by natural selection was one of the most successful and unassailable explanatory schemas in the history of science, perhaps even in the history of thought.[1] This was all the more surprising considering what the theory was designed to explain, which was, basically, all of life. If there was anything that was immediately obvious about life, it was that it existed in great profusion and almost infinite variety. There were plants, animals, microorganisms. Reptiles, mammals, amphibians, fishes. Insects, worms, algae, mosses, molds, and mushrooms. Ferns, flowers, birds, whales, and plankton. Yeasts, jellyfish, snails, sharks, diatoms, pollen grains, and quaking aspen. Even in the twenty-first century there was still no firm estimate as to the total number of species that currently existed (although the numbers ranged in the millions), much less a solid calculation of the yet larger number that had once existed and become extinct.

What unitary theory could possibly embrace all of those varied life-forms, and assert with confidence how they all came to be? Evolution by natural selection was the answer. As extraordinary as it was that any scientific framework could encompass the full domain of living things at all, what was even more astonishing about the theory of evolution was its extreme simplicity. It rested on just three key ideas: the principles of common descent, random variation, and natural selection. The descendants of a given organism were not all identical: there were variants, some of which were better fitted to survive in their environment, others of which were less so. On average, the better fit tended to leave more surviving descendants than the less fit. The same principle applied across several generations of organisms, repeatedly, and over the course of time the fitter descendants prospered ("The vigorous, the healthy, and the happy survive and multiply," Darwin wrote), while the less fit progressively died out and finally disappeared altogether. Gradually, new species arose, and old ones became extinct.

That was the theory in a nutshell. It was almost too clear, plain, and simple to be true. Nevertheless, it was. There was not the least shred of doubt that its two basic processes— variation and selection—operated in nature, and that its genetic principle of common descent was true. The question was whether those principles explained *all* of life. Did evolution account for every last nook, crag, and cranny of living things, or were some features of organisms left dangling and out of the picture, their existence unexplained by any realistic interpretation of the theory? And if evolution didn't account for those features left dangling, what did?

EVOLUTION, as a concept, did not originate with Darwin, and in fact went back to two of the earliest Greek scientist-philosophers, Anaximander (ca. 611–547 B.C.) and Empedocles (ca. fifth century B.C.), both of whom argued that later species had evolved from earlier ones. Living beings first arose in the sea, migrated onto dry land, and then evolved into progressively more complex forms of life. Humans must have evolved "from living things of another kind," said Anaximander, "since the other animals are quickly able to look for their own food, while only man requires prolonged nursing."

Even the principle of natural selection had been anticipated by the ancients. On the very first page of the *Origin of Species* (sixth edition, 1872), Darwin quotes Aristotle as having suggested that those spontaneous adaptations which were useful were retained, while those that were not were discarded: "'Wheresoever, therefore, all things together (that is all the parts of one whole) happened like as if they were made for the sake of something, these were preserved, having been appropriately constituted by an internal spontaneity; and whatsoever things were not thus constituted, perished, and still perish.' We see here the principle of natural selection shadowed forth," Darwin said.

Still, it was Charles Darwin himself who first formulated

and systematically developed, defended, and presented at great length and in scrupulous detail the full-fledged theory of evolution by natural selection.[2] Except for his voyage to Galápagos, Darwin's life story is not quite as well known as his famous theory. He was born in 1809 to a wealthy physician father and a mother who died when Darwin was eight years old. He was by inclination and practice an amateur naturalist but to please his father went to medical college, which he detested on account of the gruesome nature of surgeries performed without anesthesia.

In 1831, after taking a fallback degree in divinity from Christ's College, Cambridge, Darwin was invited to join the *Beagle* expedition, which was to be a two-year-long voyage around the world. He accepted and wound up as the ship's naturalist on a journey that in the end lasted five years. The experience changed his life.

The canonical version of the story is that Darwin's adventures in the Galápagos, a dozen or so volcanic islands off the western coast of South America, played a fundamental role in the formation of his theory. Allegedly, his observations there of "Darwin's finches," of which there were at least fourteen species on the islands, but none anywhere else, were decisive. The dissimilar beaks of these birds each seemed to be specially adapted for the different varieties of foodstuffs that predominated on the respective islands: large seeds, small seeds, leaves, cactuses, insects, and so on. Apparently, a single colonizing species had landed on the islands years before, dispersed into several distinct ecological niches, and then had gradually evolved into the different, noninterbreeding species he saw before him, each of them specially adapted to its own surroundings. A similar tale is told about the Galápagos tortoises.

More recent historians, including Stephen Jay Gould, have portrayed this account as a "romantic myth," and Gould has argued that Darwin "missed the story of the finches entirely," and also got it wrong about the tortoises. "Darwin became an evolutionist," Gould writes, not from his observations on Galápagos, but "by returning to England and immersing himself in the scientific culture of London—by arguing with colleagues, by reading and pondering (mostly in the library of the Athenaeum Club), by seeking good advice (learning from the ornithologist John Gould, for example, that those diverse Galápagos birds were all finches)," et cetera.[3]

If Darwin had a single revelatory moment, it was not on Galápagos but back in London, where in 1838, two years after the end of the *Beagle* voyage, he read *An Essay on the Principle of Population* by the British economist Thomas Malthus. In that book, Malthus claimed it to be a virtual law of nature that animal and human populations were destined to outstrip their food supplies, in consequence of which all individuals were locked into what Malthus himself, and Darwin after him, called "a struggle for existence." This struck a clear note with Darwin, who soon afterward decided that competition for scarce resources was the driving force of natural selection. "Only a few of those annually born can live to propagate their kind," he wrote much later. "What a trifling difference must often determine which shall survive, and which perish!"

As obvious as all of this is now, and indeed was even when Darwin first stated it publicly in 1859, there has hardly ever been a theory more reviled, ridiculed, and flatly rejected by a large proportion of the human race, particularly Americans, than the theory of evolution.[4] People have found it personally repugnant to imagine that they had "descended from the apes." Others see the theory as describing a mindless, soulless, and "materialist" world. And then there are those who for religious reasons believe that species, the world, and everything in it were the product of a supernatural creator who miraculously breathed life into inanimate objects.

But there has scarcely been a theory with more factual evidence behind it. The current scientifically accepted theory of evolution, known as the Modern Synthesis (also called Neo-Darwinism), combines the Darwinian principles of common descent, random modification, and natural selection with Mendel's laws of genetics and with the mechanisms of inheritance as explained by DNA and all the related substructures of molecular biology. Proof of the universal common descent of all organisms, for example, is provided by the fact that the living cells of all earthly creatures share the same basic methods, means, and modes of operation: they express their genetic information in nucleic acids, use the same genetic code to translate gene sequences into amino acids, and (with some exceptions in the case of plants) make use of the same twenty amino acids as the building blocks of proteins. Since all of these features are essentially arbitrary, the fact that they are universal among all the life-forms of earth is strong evidence of their common ancestry.

Further, many organic molecules have a "handedness," or chirality, meaning that the molecule cannot be superimposed upon its mirror image, for the two are asymmetrical. In chemistry, the left-hand version of a molecule is known as its L form (for *levo*, Latin for left), the right-hand version as its D form (for *dextro*, Latin for right): lactic acid, for example, exists as both L-lactic acid and D-lactic acid. Throughout nature, however, the cells of living organisms utilize predominately L-versions of basic organic molecules. The amino acids that are used to make proteins, for example, are all L-amino acids, none of them D-versions.

"The first great unifying principle of biochemistry," Francis Crick has said, "is that the key molecules have the same hand in all organisms." Uniformity of biochemistry, however, implies commonality of descent.

Then there is the fact that when molecular biologists compare the genomes of closely related species, their gene sequences are found to be remarkably alike: for instance, there is a 95 percent similarity between the genomes of humans and chimpanzees. Further, there is the evidence of vestigial structures: the human appendix and coccyx, for example, are relics of functional structures in ancestral animal species. There is the evidence from embryology, which shows that an early human embryo is morphologically similar to those of other mammals such as the dog, cow, and mouse, and in its earliest stages is also similar to the embryos of reptiles, amphibians, and fishes. At one point in its development, the human embryo even possesses gill-like slits, and indeed exactly the same number of them (four) as the embryos of fish, salamanders, tortoises, chickens, and pigs, among other animals.

Taken together, all this, plus other substantial evidence— from comparative morphology (which reveals anatomical similarities among related species), from the fossil record (which shows strata of more highly developed organisms layered on top of older and more primitive ones), from the routine discovery of so-called missing links (which fill in gaps in the fossil record)—all of these facts and more make it difficult to deny the contention that "higher" species of organisms have descended from "lower" ones.

Recognizing this truth nevertheless does not imply that there are no problems or open questions connected with the theory of evolution. In fact the theory has a fair share of them.

CONSIDER, for example, the phenomenon of altruism, or self-sacrificial behavior, a trait that is not limited to humans. Some birds, for example, are known to emit self-endangering cries in order to warn their flock of approaching predators. Far from being adaptive, such behavior appears to be positively destructive inasmuch as the systematic practice of it would ultimately lead to the death of self-sacrificing individuals. In theory, then, altruistic behavior ought to vanish over time. Instead, it persists.

Several evolutionary solutions have been offered to this problem, but none seems wholly satisfactory. Darwin himself, when confronted with the task of explaining altruistic behavior among humans, argued that although the behavior was harmful to the individual, it was nevertheless beneficial to the group of which he or she was a member—the tribe, for example. But that solution, which amounted to a theory of group or species selection, forced Darwin to abandon one of the basic assumptions of his theory, which was that the unit of selection is the individual, not the group. "Natural selection acts only by the accumulation of slight modifications of structure or instinct, each profitable to the individual under its conditions of life," he said in one of many similar passages in the *Origin of Species.*

Another solution proposes that altruistic individuals act with the expectation that others will in turn act altruistically toward them—the theory of "reciprocal altruism" (or what might be better termed "selfish altruism"). This, too, happens in the animal world. Vampire bats, for example, cooperate by sharing food with other bats instead of hoarding as much as they can for themselves. They even keep track of which other bats share their food with them, and thus practice a "you scratch my back, I'll scratch yours" philosophy.

That philosophy, however, fails to explain self-sacrificing acts that benefit complete strangers when there is little or no hope of the stranger ever returning the favor—for example, when a passenger gives up his seat to an elderly person on a bus or when a soldier throws himself on a bomb that's about to explode in order to save the lives of people he doesn't know in the least. A contemporary range of evolutionary add-ons— including the new-wave fields of cultural evolution, evolutionary psychology, and sociobiology—attempt in part to account for such behaviors, but these disciplines have attracted critics and engendered bitter controversies of their own, and so cannot be regarded as having solved the altruism problem.[5]

The problem of altruism by no means overturns the theory of evolution by natural selection, but it exposes traits that the theory by itself doesn't seem to cover. On the other hand, Darwin himself never claimed that his theory was the sole explanation of all the structures and behaviors of living organisms. In the *Origin of Species* he stated twice that selection doesn't explain everything, saying at the end of the book's introduction: "I am convinced that Natural Selection has been the most important, but not the exclusive means of modification." He repeated the statement almost verbatim in the closing section of later editions, in which he said: "As my conclusions have lately been much misrepresented, and it has been stated that I attribute the modification of species exclusively to natural selection, I may be permitted to remark that in the first edition of this work, and subsequently, I placed in a most conspicuous position—namely at the close of the Introduction—the following words: 'I am convinced that natural selection has been the main but not the exclusive means of modification.' This has been to no avail."

Darwin, in a sense, was not a Darwinist, if Darwinism is taken to mean that every last crumb and pebble of the biological universe is the exclusive product of natural selection—or, as the geneticist Theodosius Dobzhansky wrote in 1973, that "nothing in biology makes sense except in the light of evolution." That is simply false. One of the more memorable examples of its falsity was given by the evolutionary biologist George C. Williams in his 1966 book *Adaptation and Natural Selection*: "Consider a flying fish that has just left the water to undertake an aerial flight." Is its return to the water a product of natural selection? "Certainly not; we need not invoke the principle of adaptation here. The purely physical principle of gravitation adequately explains why the fish, having gone up, eventually comes down."

Indeed, even within the Modern Synthesis there are ample controversies among evolutionary theorists. One of the foremost concerns the unit of selection, the physical object that the process of selection operates on. Is it the gene, the cell, the individual, the species, a combination of them all, an entire population, or something else?

Darwin held that selection operated on the individual organism. Much later, in light of Watson and Crick's elucidation of DNA's structure and Marshall Nirenberg's deciphering of the genetic code, the view became widespread that the unit of selection was the gene, a notion popularized by the British evolutionist Richard Dawkins in his 1976 book *The Selfish Gene*. But that hardly settled the issue.

The years 2001 and 2002 proved to be the high-water mark in the unit-of-selection debate. In this short span of time, the two top evolutionary theorists in the United States, both of them Harvard professors, each published books on evolution, and their answers to the unit-of-selection question did not agree. One was Stephen Jay Gould's magnum opus, *The Structure of Evolutionary Theory*, a dense 1,343-page text in which he laid out his final thoughts on the subject. (Gould died in 2002, as the book was being published.) In it he expressed his view that the unit of selection was not the gene, the individual, or the species, but rather "a nested hierarchy of biological individuals (genes, cell lineages, organisms, demes [a local population of interbreeding individuals], species, clades [a group of species sharing a common ancestor])."

The other book was Ernst Mayr's *What Evolution Is*.

While Gould was regarded as "America's evolutionist laureate" by some (and in 2001 had been named by the Library of Congress as one of America's eighty-three Living Legends), Mayr was revered by many as the grand old man of evolutionary biology. He had been born on July 5, 1904, which meant that at the time his book was published in 2001 he was ninety-seven years old.[6] An entire book collection at Harvard's Museum of Comparative Zoology, the Ernst Mayr Library, had been named after him. His own book *What Evolution Is* included two appendices, one of which was titled "Short Answers to Frequently Asked Questions About Evolution." Question 18 was: "What is the object of natural selection?" to which Mayr answered: "Why has there been so much controversy about the object of selection?" as if this were a minor detail of peripheral interest. "At the time of the evolutionary synthesis, the geneticists believed that it was the gene, whereas the naturalists believed it was the individual, as Darwin had always believed. Forty years of analysis have finally made it clear that the gene as such could never be the direct target of selection. However, in addition to the individual, a group can also be the target of selection . . . Finally, gametes are also directly exposed to selection . . ."

That two of the most authoritative recent books in the genre, both of them written almost 150 years after the first publication of the *Origin of Species*, could disagree about so fundamental a matter as the unit of selection was mildly incredible. It suggested that the theory of evolution as it stood was not quite the last word on the subject of how biological organisms came to possess their full array of physical attributes and behavioral traits.

MANY RECENT EVOLUTIONARY THEORISTS nevertheless continued to interpret virtually every facet of biological life in strictly adaptationist terms. It was as if they were forcing biological phenomena to fit the theory rather than altering the theory to fit the facts.

In the late 1970s Stephen Jay Gould had a revelatory moment of his own, one in which he saw a way to combat what he regarded as the "Panglossian" tendency of evolutionists to interpret all biological phenomena as adaptations for the best in this best of all biological worlds. While attending a scientific conference in Venice, he, like any other tourist, made the compulsory pilgrimage to San Marco basilica, one of the emblematic architectural masterpieces of the city. He stood under the largest and most central of the basilica's five domes and looked up. What he saw there would add a new wrinkle, and some new language, to the Modern Synthesis.

He saw *spandrels*.

Spandrels are architectural features created when a hemispherical dome is supported by a set of four rounded arches. Physically, they are tapering triangular expanses that stretch down from the bottom edge of the dome to the tops of the arches below—"spaces left over."

What could they possibly have to do with evolution?

Gould answered the question in 1979 when he and his Harvard colleague Richard Lewontin wrote a paper called "The Spandrels of San Marco and the Panglossian Paradigm: A Critique of the Adaptationist Programme." Published in the *Proceedings of the Royal Society of London* and reprinted

many times since, it was to become famous among evolutionary biologists and inspired a series of rebuttals, spin-offs, and
even parodies, two of which were entitled "The Scandals of
San Marco" and "The Spaniels of St. Marx."[7] The paper was
their attempt to correct a theory of evolution that they saw as
having gone slightly berserk, its principles being applied so
widely that adaptation "becomes the primary cause of nearly
all organic form, function, and behavior." They identified
such a program as Panglossian because of the manner in
which it assumed "the near omnipotence of natural selection
in forging organic design and fashioning the best of all possible worlds . . . Any suboptimality of a part is explained as its
contribution to the best possible design of the whole. The notion that suboptimality might represent anything other than
natural selection is usually not entertained."

Gould and Lewontin offered several examples from the
literature in which adaptationists gave evolutionary explanations of structures or behaviors in cases where equally or more
plausible nonadaptationist explanations were clearly available. One example concerned aggression in mountain blue-
birds: an experimenter placed a stuffed male bluebird near
the nests of two bluebird couples while the males were out
foraging. The experimenter did this both before and after the
female laid her eggs, and then observed results. Before the females laid their eggs, the returning male birds attacked the
stuffed male as if it were a threat. After the females laid their
eggs, however, the attacks lessened in frequency. According
to the experimenter, this made evolutionary sense because
the males would be naturally more protective of the female
before she laid her eggs than afterward, when the male could

be sure that his precious selfish genes were safely lodged inside the eggs.

"But what about an obvious alternative?" Gould and Lewontin asked. When the males return, they recognize the stuffed bluebird as "the same phony they saw before," realize there's no danger from it, and therefore curtail their attacks.[8]

Not everything is an adaptation, Gould and Lewontin all but shouted: spandrels were not. Spandrels were not designed and inserted by the architect to answer a specific need (for example, as attractive spaces to display mosaics). Just the opposite: "The spaces arise as a necessary by-product of fan vaulting." In the same way, some parts of some organisms, and some behaviors, are not adaptations, either; rather, they are necessary by-products of structures that are adaptations. They are *nonadaptive spandrels*.

Later, in *The Structure of Evolutionary Theory* (and elsewhere), Gould offered several examples of nonadaptive spandrels in biology: *the whiteness of bones* ("Bones are made of calcite and apatite for adaptive reasons, but bones are also white because the chemistry of these compounds so dictates"); *the human brain* ("Natural selection didn't build our brains to write or read, that's for sure, because we didn't do those things for so long . . . Most of its modes of working don't have to be direct results of natural selection for its specific attainments").[9]

And then there was the example of *male nipples*, which had puzzled biologists since Buffon first raised the question as a serious scientific issue in the 1700s. A few adaptationist explanations had been offered in the years since: maybe the males could suckle infants in certain circumstances. Maybe they once did, and current nipples were vestigial. "But the probable resolution," said Gould, "based on a quite different (albeit simple) perspective, requires the concept of nonadaptive spandrels: males probably grow nipples because females need them for an evident purpose, and many aspects of [embryological] development follow a single pathway. So females grow nipples as adaptations for suckling, and males grow smaller and unused nipples as a spandrel based upon the value of single development channels."

In this way the concept of spandrels entered the lexicon and literature of evolutionary biology.[10]

WITH THE THEORY of evolution's domain thus newly, if only slightly, restricted, the question arose of whether there was some underlying systematic mechanism that could explain the existence of spandrels and other biological features for which adequate adaptationist explanations did not seem feasible or realistic. Stu Kauffman, who at the fiftieth-anniversary commemoration of Schrödinger's book had questioned both DNA and evolution as all-powerful explanatory schemas, chimed in once more with his pet process, self-organization. By means of complex autocatalytic biochemical reactions, he suggested, certain features of organisms spontaneously organized themselves into existence.

In no way did Kauffman deny the operation of natural selection; he merely argued that it had to be supplemented by something else. "Natural selection is always acting," he said. "The natural history of life is some form of marriage between self-organization and selection."

Stephen Gould, for one, liked this idea, saying that Kauffman was "groping towards something important." Still, the question remained open, a project for further research.

But there were far more novel, original, and radical ideas than self-organization on the horizon as alternative explanations for otherwise unaccounted-for biological phenomena. Most revolutionary of them all were the mechanisms proposed by Stephen Wolfram in his 1,200-page megabook, *A New Kind of Science*, published in 2004.

Wolfram was not an evolutionary theorist, or even a biologist. Trained as a physicist, he was the recipient of a MacArthur "genius" award (as were both Gould and Kauffman), and more lately was the creator of the fabulously successful scientific software system known as Mathematica. That was mere prelude; with the proceeds from Mathematica, Wolfram withdrew from the world to create what he regarded as a new kind of science.

His book of that title, which he referred to by its acronym, NKS, was, as he explained in the preface, the product of nearly twenty years of work, more than a hundred million computer keystrokes, and more than a hundred mouse miles. During all that time he had peered down into the maw of nature and saw there the workings of *cellular automata*.

Cellular automata (CA) were computerlike programs, simple mathematical rules that when applied repeatedly were capable of generating patterns or behaviors of the most devastating complexity. Wolfram decided, after years of simulating the outputs of these programs, that cellular automata were so varied and fertile that they could produce virtually all of the complexity found in nature, including much of it that was found in biology. The world, in his view, was something like a gigantic computer running on a kind of natural cellular automata operating system, and the physical universe and most of the things in it were products of implementing the operating system's total set of rules.

Wolfram, it turned out, took a fairly dim view of evolution. Although he did not deny that it operated in nature, he claimed that what it could achieve was rather meager.

"While natural selection is often touted as a force of almost arbitrary power, I have increasingly come to believe that its power is remarkably limited," he wrote. In fact, natural selection by itself might not produce structures of great complexity. One piece of evidence for this, he said, was that relatively complex features were found even in the very earliest fossilized organisms. Complexity, therefore, must have been comparatively easy to achieve, and might well be a natural outgrowth of CA programs, rather than of evolution. Another piece of evidence concerned the pigmentation patterns of organisms, which varied from the very simple to the highly complex. Wolfram's explanation for this was that some of the underlying CA rules produced simple patterns while others did not, just as he had observed countless times in simulations.

Pigmentation patterns indeed constituted what was perhaps Wolfram's greatest strength when it came to explaining biological phenomena. NKS was brimming with pictures of highly patterned biological organisms coupled with abstract designs that had been generated by various discrete cellular automata rules. The best example of a parallelism between the two concerned the pigmentation patterns on certain mollusks, some of which were remarkably similar to the patterns produced by specific, known CA programs. Those pigmentation patterns, Wolfram argued, were unlikely to have been produced by natural selection, the reason being that "in many species of mollusks the patterns on their shells—both simple and complex—are completely hidden by an opaque skin throughout the life of the animal, and so presumably cannot possibly have been determined by any careful process of optimization or natural selection."

In Wolfram's view, the similarity between the mollusk patterns and the output of CA programs was not coincidental.

"A mollusk shell, like a one-dimensional cellular automaton, in effect grows one line at a time, with the new shell material being produced by a lip of soft tissue at the edge of the animal inside the shell . . . And given this, the simplest hypothesis in a sense is that the new state of the element is determined from the previous state of its neighbors—just as in a one-dimensional cellular automaton."

It would be putting things mildly to say that Wolfram's claims, his book, and he himself, were highly controversial.[11]

At the farthest extreme, finally, one could even question whether the ability to evolve was a necessary condition of an entity's being considered alive. Since many organisms do not leave offspring, whether out of biological necessity (e.g., sterile hybrids), choice, or sheer accident, these nonparents have no descendants, adaptive or nonadaptive. But it would be extremely odd to regard them as not alive on that account. One could even imagine a species of organism that did not reproduce, and therefore could not evolve: they simply popped into existence through some type of self-organizing agency, lived for a while and perhaps grew in size, and then died.

Whether this birth-growth-death process was continuously repeated or was a onetime phenomenon, the organism in question would seem to be at least minimally alive in either case.

Such science-fictional cases aside, what the primary evolutionary disputes—over the problem of altruism, the unit of selection, spandrels, not even to mention further arguments over punctuated equilibrium and the like—went to show was that the Modern Synthesis is not exactly the final, finished, and polished theory that its defenders sometimes portray it as being. Other scientific theories have been accepted early on and have remained virtually unchanged over the years: Einstein's theories of special and general relativity, for example, or the essentials of quantum mechanics. As a theory, evolution has not fared nearly as well. The platitude seems unavoidable that the theory of evolution is itself evolving and may continue to do so for some time to come.

Eight

The Twilight Zone

IT BECAME APPARENT in the last few decades that what unaided nature could achieve over aeons through evolution by natural selection, scientists could accomplish at a much faster rate by manipulating the genes themselves. There were two quite distinct ways of doing this, however: an indirect, "natural" method and a direct, "artificial" one. The "natural" method, employed throughout most of human history, including the present, operated on whole organisms and was practiced in the barnyard, greenhouse, farm, or vineyards — wherever people crossbred animals or plants for specific desired traits. That kind of indirect genetic modification had been going on for such a long period on such a vast scale and was performed on such a wide variety of plants and animals, including dogs, horses, cows, flowers, vegetables, fruits, and other flora and fauna, that few people ever gave it a second thought. Still, primitive and rudimentary though it was, this was unquestionably a form of genetic engineering inasmuch as the end products were genetically different from their ancestors and the farmer, plant grower, or livestock breeder had achieved his results deliberately. The fact that the practitioners of these somewhat arcane pursuits often enough didn't know the first thing about genes or have any idea of how their biological or botanical improvements were made at the molecular level didn't mean that they weren't in fact reworking the hereditary materials of the organisms in question, recombining molecular structures and thereby creating new and enhanced genes.

Nobody ever objected to doing things the "natural" way, and so that method of "interfering with the natural order" was seen as benign and even boring, or in the case of hybrid flowers, faster horses, or better-looking dogs was viewed as a positive benefit to humanity. This was true despite the fact that nonspecific interbreeding had the potential to create "monsters," or at least some fairly strange-looking beasts. (For that matter, natural births in the wild, and even in humans, also occasionally resulted in "monstrous" or grotesquely abnormal offspring that were usually not viable.)

The "artificial" method of producing improved varieties of organisms, by contrast, worked by manipulating the genes directly. This method required more than a barnyard and a couple of animals; it required a substantial outlay of resources in the form of laboratories, lab glassware, hardware, software, machinery, equipment, and instrumentation; stocks of chemicals, enzymes, nutrients, and other media, along with intimate biological knowledge of the organism under investigation. In the canonical genetic engineering (or recombinant DNA) experiment, a given length of DNA was removed from a chromosome (or from a plasmid, a circular stretch of DNA that replicated independently of a cell's genome) by the use of a restriction enzyme, a protein that had the ability to cut DNA at specific sites. A new gene was inserted where the original one had been and then spliced into the chromosome (or plasmid) by means of a ligase, an enzyme that acted as a sort of molecular glue, and which smoothly interpolated the new DNA fragment into that of the existing gene. The outcome was an "engineered," or directly modified, gene, and a genetically modified organism.

Genetic engineering in this sense was pioneered by Paul Berg, Stanley Cohen, and Herbert Boyer in the biology labs of Stanford University and the University of California, San Francisco in 1973. Because the technique worked by taking known stretches of a preexisting organism's DNA and combining them with select DNA segments taken from another organism, it was an essentially conservative form of modifying life. (Indeed, specialized versions of the procedure occurred all the time in nature. In one version, called conjugation, bacteria exchanged genes across a protein bridge that was in effect a tube for the transmission of DNA from one bacterial cell to another. A second version, called transduction, occurred when a bacteriophage virus took part of the DNA out of one bacterium and transferred it into a recipient. These were routine and entirely natural processes.) Nevertheless, because "artificial" genetic engineering took place in the laboratory, gene-splicing technology was viewed by some as being "unnatural," Frankensteinish, and a potential threat to humanity. It was for such reasons that in the summer of 1976 the city council of Cambridge, Massachusetts, banned certain recombinant DNA experiments from taking place within the city limits for a period of three months.

Still, cool rationality eventually reasserted itself. After all, genetic engineering was motivated, at least in the beginning, by narrowly practical aims, such as getting bacteria to produce valuable substances such as drugs, vaccines, and so on. Despite the fact that it was a direct molecular interference with the natural order, the general public ultimately accepted gene-splicing as just another one of the modern miracles provided by science, and it soon became peacefully integrated with the rest of the burgeoning universe of advanced technology. In 1982, the Food and Drug Administration approved the first genetically engineered biological substance for use in human medicine, insulin produced by genetically altered *E. coli* bacteria. Shortly thereafter, the FDA approved genetically engineered human growth hormone and a hepatitis vaccine for use in humans. By the end of the 1980s, genetic engineering had become a successful, practical, and entirely innocuous business enterprise.

But it was not long before scientists were using those same genetic techniques to reengineer life in more disturbing ways, producing "chimeras," for example, transgenic species that were blends of two different organisms. For a while scientists seemed to be in the grip of a lurid fascination with combining the genes of naturally light-emitting organisms with those that weren't, creating, for example, a tobacco plant that glowed in the dark. This wonder shrub had been invented in 1986 by introducing into the tobacco's DNA the luciferase gene of the firefly, the enzyme responsible for its bioluminescence. Ostensibly, the purpose of the exercise was to create a "reporting mechanism," a diagnostic tool that signaled when the fusion of the new gene into the old one had been successfully accomplished. In this case, since the plant lit up like a lightning bug, it had been.

Still, that was nothing. Soon some even stranger luminescent creatures were growing, swimming, or crawling around in the lab: bacteria that flashed on and off like Christmas tree lights, fish and mice that emitted a dim radiance, plus a fluorescent albino rabbit named Alba. In 1999, the Chicago artist Eduardo Kac proposed the creation of GFP K-9 ("green fluorescent protein canine"), a dog to be enhanced with a gene that coded for the fluorescent protein from a luminescent jellyfish that would cause the dog to glow under ultraviolet light, the resultant lustrous animal to be presented to the art world as a "transgenic art object."

Weird as they were, all these creatures had been produced by using nature's own genes, molecular structures as natural as raindrops, only mixed together in novel ways. But what if you fused a natural gene not with *another* natural gene but, rather, with an *artificial* one that came not from a bacterium, plant, or animal, but from a machine?

THE FIRST COMMERCIALLY SUCCESSFUL "gene machine," or DNA synthesizer, had been invented in the early 1980s by Leroy Hood and Mike Hunkapiller of Applied Biosystems, in Foster City, California, as a molecular biology research tool. The company's Model 380A DNA Synthesizer, which came out in June 1983 and which was priced at $42,500, was a desktop unit with an alphanumeric keypad into which you could type any desired sequence of DNA bases, press <ENTER>, and watch as the device deposited millions of copies of exactly that sequence into a tiny white vial, ready for use, only minutes later.

Plainly, the DNA synthesizer was a mad scientist's dream machine. With it, you could genetically tinker with an organism in ways never before possible, endowing an animal with some offbeat genes of your own design. For although it had been made mechanically, using bottled reagents as raw materials, the final result was not some sort of fake or pretend DNA; it was "real" DNA, functionally indistinguishable from the same sequence as it appeared in the living organism. And since it was entirely genuine DNA, a given stretch of it as synthesized by the gene machine could be inserted into the cells of a living organism exactly in the manner of a conventional genetic engineering experiment. Whether your homegrown DNA was operational when dropped into the genes of a given organism was another matter, but that was what experiments were for.

Thus was born the prospect of a new kind of biology—synthetic biology—one in which an organism could have some of its original, natural DNA replaced by one or more DNA sequences of human devising.[1] By 2002, scientists had actually created an entire microorganism out of synthetic gene sequences alone. The first organic structure to be so created, however, was a rather unexpected one: the polio virus. Moreover, it was constructed by the scientist who had first decoded the genetic sequence of the natural virus, Eckard

Wimmer. Wimmer was a microbiologist at the State University of New York at Stony Brook, and in 1981 he and a substantial group of coworkers announced in the pages of *Nature*: "The primary structure of the poliovirus has been determined. The RNA molecule is 7,433 nucleotides long."

Some twenty years later Wimmer, this time with a new but smaller team of assistants, had re-created that virus from the very genome sequence that he himself had discovered. This was not merely a matter of keying in all the 7,433 base-pair sequences into a gene machine, however, the reason being that DNA synthesizers generally produced only short stretches of DNA, snippets known as "oligonucleotides," structures some 400 to 600 bases in length. The problem was in putting those snippets together in the right order. Nevertheless, following a protocol that he downloaded from the Internet and using gene sequences purchased from a mail-order supplier of custom-made DNA, Wimmer and his crew were finally successful in the *de novo* synthesis of the original infectious poliovirus. To make sure that it was the genuine item, they injected samples of the virus into transgenic CD 155 mice, which were specially bred to be susceptible to polio, and found that the synthetic virus caused paralytic poliomyelitis in the rodents, just as the natural virus did.

In its own way, the experiment was a tour de force. "Our results show that it is possible to synthesize an infectious agent by in vitro chemical-biological means solely by following instructions from a written sequence," Wimmer wrote in the pages of *Science*.

That itself would have been an example of creating life from scratch with the aid of an instruction manual and a few machines except for the fact that a virus was not a living thing, but rather only a string of dead chemicals inside a protein coating. Still, the techniques involved, like any others in the sciences, had the potential to be refined, accelerated, and made more powerful, bringing the *de novo* creation of life ever closer to realization.

A year after the synthetic polio virus, a team composed of Craig Venter, Hamilton Smith, and associates had created another virus from scratch, the phiX174 virus. PhiX was not infectious to humans and was simpler than the polio virus, for it consisted of only 5,386 base pairs. Venter and Smith's claim to fame was that they created the phiX virus not over a period of months but in the space of two weeks. They designed their "oligos," ordered them from Integrated DNA Technologies, of Coralville, Iowa, ligated them (glued them together) in the proper order, and after performing several routine lab assays, amplifications, purifications, and tests, announced they had successfully brought the virus into being from mere information on a page. "We have demonstrated the rapid, accurate synthesis of a large DNA molecule based only on its published genetic code," they wrote in their technical report on the project.

The authors were by no means oblivious to the broader implications of what they had wrought: "The creation of life in the laboratory can now be considered a realistic possibility," they said.

THE NEXT YEAR, 2004, was to be the start of a banner era in the realm of synthetic biology. For one thing, it was in the summer of 2004 that a couple of biological engineers at MIT, Drew Endy and Tom Knight, set up a Standard Registry of Biological Parts, a supply bin of microbe components. Endy had come to synthetic biology from engineering, Knight from the ever-promising, ever-disappointing world of artificial intelligence. After a while both of them had realized that biological organisms were malleable enough and their molecular components interchangeable enough that they could be designed, engineered, and put together from off-the-shelf "factory" parts, exactly after the manner of cars, washing machines, or toasters. The ultimate goal of such an approach, Endy said in a bold quote, would be to "reimplement life in a manner of our own choosing."

That was not as wild as it sounded. Life already existed in practically an endless variety of forms, and many organisms could be altered easily enough so as to give them additional properties, processes, and abilities. Genetic engineering already got bacteria to make insulin, human growth hormone, tissue plasminogen activator, erythropoietin, vaccines, and other materia medica. Synthetic biology ought to be able to take the process one step further, allowing you not only to reprogram bacteria but actually to build new types of them to specifications announced in advance. All you needed was a ready supply of the necessary ingredients and the knowledge of how to put them together.

MIT's Standard Registry of Biological Parts was to be a repository of biological bits and pieces, snippets of DNA that encoded for known structures and/or functions, so that you could know beforehand what would most probably happen if you put one or more of them into your experimental organism. These molecular components would be reliable, dependable as bricks, and in fact Endy and Knight went so far as to call these biological parts "BioBricks."

One of the first to be accessioned into their registry was—no surprise—a gene for making organisms glow in the dark. In January of 2004, Endy had offered an MIT Independent Activities course in which students got *E. coli* bacteria to fluoresce, in what had by now become almost a rite of passage for any aspiring genetic engineer. But this being MIT, and both Endy and the students deeply proud engineering and technology geeks, they were not content to use nature's own GFP (green fluorescent protein) gene as found in jellyfish. Instead, they would use a gene of their own design and have it custom-built by an industrial gene supplier. Accordingly, they placed an order with Blue Heron Biotechnology in Bothell, Washington, a company whose web page announced that it could "synthesize any gene regardless of sequence, complexity, or size with 100% accuracy," and whose advertising slogan was "You get the gene you want." (In 2006, the company's going rate for *de novo* gene synthesis was $1.60 per base pair.)

It took about a year for the group to install the synthetic gene in *E. coli* and get it working properly so as to light up the organism as planned, but it finally happened, and the synthetic fluorescence BioBrick was added to the inventory. By the end of 2004, MIT's Standard Registry of Biological Parts contained 140 BioBricks, all of them stored either in small vials sealed with yellow stoppers and placed in freezer boxes or alternatively deposited into multiwell plates. There were "promoter" BioBricks, gene sequences that initiated the transcription of DNA into mRNA; there were "terminator" Bio-

Bricks, "inverter" BioBricks, and so on. (The list of BioBricks current at any given time could be viewed at the Registry's public website: parts.mit.edu.)

That same year, 2004, the Bill and Melinda Gates Foundation announced a $42.5 million grant in support of a synthetic biology project whose goal was to rewire *E. coli* bacteria so that they would produce the antimalaria drug artemisinin. Prior to that time, artemisinin, which as a traditional Asian remedy had been used for two thousand years to treat a variety of ailments, could be found only in minute quantities in wormwood, a plant that flourished primarily in the mangrove swamps of China and Vietnam. Extracting the drug was an expensive process, however, and one that was further complicated by the politics, economic consequences, and ecological correctness of harvesting the plant in developing countries. But Jay Keasling, of the University of California, Berkeley, and the Lawrence Livermore National Laboratory, had a plan to combine genes from wormwood, *E. coli*, and yeast, causing the resulting microbe to generate an artemisinin precursor chemical that could then be turned toward manufacturing the drug cheaply.

Suddenly, synthetic biology was more than just an interesting academic exercise. It, too, was a business enterprise. But for all its rhetoric about "reimplementing life in a manner of our own choosing," and for all the practical applications it offered (which ranged from getting bacteria to produce drugs, new-wave fuels such as hydrogen for use in cars, and other complex chemicals, to the invention of organisms that would report the presence of TNT, sarin, and other chemical, biological, or explosive weapons, or even to digest quantities of VX, the U.S. Army's lethal nerve agent), the question remained as to whether synthetic biology was really anything more than glorified genetic engineering. Did it, in the end, shed any light at all on what it meant to be alive? In truth, it was not so much "synthetic" biology as it was "assisted" biology: it started with conventional biological organisms and merely added something new to them, or subtracted something else, or otherwise modified an already existing life-form. It was, in other words, an attempt not so much to create new life as to modify old life in new ways.

For that matter, analogous questions could be asked regarding the creation of viruses. Scientists had decoded their gene sequences, invented machines that could synthesize them, and then put those same sequences back together again in the order they had been in to begin with. Was this really all that great an accomplishment? Francis Collins, the head of the American Human Genome Project, didn't think so. "This was completely a no-brainer," he said in a 2005 interview. "I think a lot of people thought, 'Well, what's the big deal? Why is that so exciting? So you synthesized a virus from scratch? Of course!'"

STILL, no such skeptical doubts attended a highly original development that took place at the Rockefeller University in New York in the spring of 2004. If there was any advance in synthetic biology that took biologists into radically new territory—into the twilight zone between natural and artificial life—it was this one.

The experiment was reported in the *Proceedings of the*

National Academy of Sciences under the title "A Vesicle Bioreactor as a Step Toward an Artificial Cell Assembly." Like the protocell scientists of Venice, Bonn, Los Alamos, and elsewhere, the Rockefeller University researchers, Vincent Noireaux and Albert Libchaber, had taken as their goal the creation of an artificial cell. And like the protocell scientists, they had adopted a vesicle as their medium of choice for the cell's corpus, the body that would eventually spring into life and thereafter function as an artificial living entity.

There the similarities ended, however, for where the Four Protocell Musketeers wanted to construct their cell from nonbiological ingredients, Noireaux and Libchaber freely borrowed from different realms of natural life. They constructed their vesicle, a spherical molecular structure with an open interior, out of phospholipids—in this case fat molecules taken from egg whites. To make the vesicle membrane permeable, they perforated it with a toxin taken from the bacteria *Staphylococcus aureus*, a substance that made tiny holes in the cell walls. Finally, they filled the vesicles with a cell-free extract taken from that omnipresent workhorse bacterium *E. coli*.

Cell-free extracts were nothing new in biology. They were useful because they constituted a biological system that contained enough of a natural cell's viable matter to support certain basic life functions—protein synthesis being the primary example—without the extraneous encumbrances of cell walls, nuclei, and so forth. As derivatives of living microorganisms, cell-free extracts were already on the borderline between life and nonlife: they performed some of the functions of living systems without having all the trappings of living systems; in addition, they could not replicate themselves. Nevertheless, they were in some extended and attenuated sense a form of *acellular life*, something on the order of a brain-dead human being.

In the old days, cell-free systems had been prepared manually by the researcher, often by putting a quantity of *E. coli* into a glass homogenizer, which was a test tube with a close-fitting piston that spun around inside it. The revolving disk would rapidly convert a supply of discrete and intact cells into a liquid suspension whose various components could then be separated out by centrifugation. Nowadays, cell-free extracts were available commercially from firms such as Roche Applied Science and could be ordered over the Internet. As Roche described one such system, "An optimized *E. coli* lysate [the contents of ruptured cells] contains tRNAs, ribosomes, enzymes, factors, and other necessary components required for highly efficient translation." In fact, it was precisely a Roche RTS 500 cell-free *E. coli* system that, in 2004, Noireaux and Libchaber used in their artificial cell experiment.

The important thing is what they did with the substance once they received it. They took this semiliving, semidead cell-free blend and inserted it into their homemade, prefabricated vesicles. In other words, they placed what was normally inside a cell *back into a cell*—but an artificial cell of their own creation. After filling it with the cell-free extract, the experimenters called the object a "vesicle bioreactor," a "cell-like bioreactor," and "a synthetic cell."

By any measure, this was a highly unusual specimen: a vaguely spherical object whose membrane consisted of egg-white fats and whose interior consisted of *E. coli* cytoplasm loaded with large quantities of the biological mechanisms normally found in living cells. Those same mechanisms now enabled the synthetic cell to do one of the main things that ordinary cells did, which was to express proteins. This means that their cobbled-together cell-like bioreactor was able to perform the workaday cellular functions of transcription (the conversion of DNA into messenger RNA) and translation (the production of an amino acid sequence that ultimately resulted in a protein), two of the most basic and characteristic functions that occurred inside conventional cells.

Most surprising of all, perhaps, was that the cell-free material supported the transcription and translation functions for a longer period when it had been encapsulated in vesicles and fed nutrients than when it remained outside the vesicle in bulk form. "Whereas in bulk solution expression . . . stops after 2 h," the authors said in their technical report, "inside the vesicle permeability of the membrane to the feeding solution prolongs the expression for up to 5 h." Furthermore, when Noireaux and Libchaber adjusted the experimental conditions for maximum nutrient intake, "the reactor can then sustain expression for up to 4 days."

And what was the protein being transcribed, translated, and expressed inside this remarkable cell-like bioreactor? None other than the canonical reporting molecule of genetic engineering: GFP, green fluorescent protein, the very same substance that lit up the glowing tobacco plant, endless batches of E. coli, mice, and even Alba, the albino rabbit that glowed in the dark. The artificial cell created out of parts of formerly living entities radiated light like a ghost.

If this was not the biological twilight zone—some sort of bizarre transitional stage between life and nonlife—what was?

Nine

The Synthetic Cell Turing Test

BY MAY 2006, a year after my visits to ProtoLife s.r.l. at Parco Vega in Marghera, the ECLT at the Palazzo Giovanelli in Venice, and the Fraunhofer Institute Schloss Birlinghoven near Bonn, a few changes had taken place across the proto-cell landscapes of Europe and America.

John McCaskill, the leader of the PACE consortium, who had been at the Fraunhofer Institute since 2003, had moved back to the Ruhr University Bochum. Some two years after he'd arrived in Bonn, the Schloss Birlinghoven branch of the Fraunhofer Institute had modified its institutional orientation to focus more on applied, maturing technologies, whereas McCaskill's project was decidedly still in the basic, pure-research mode, and so there was no longer a strict consan-guinity of interests between the two. The Fraunhofer thus lost the principal investigator, scientific team, and one of the more prestigious science grants to emerge from the European Union in recent years.

Back at Ruhr University Bochum, McCaskill and his group continued to develop and optimize their electronically controlled, regulated microfluidic systems for protocells—which is a long way of saying that they were improving their artificial life-support systems for a so-called "complemented protocell," if and when it arrived on the scene.

"'Complemented,'" McCaskill said, "means supported strongly by a microfluidic system, so much so that many scientists, including myself, will not want to credit this as the first instance of an artificial chemical cell, but as a hybrid entity that will have enough scientific interest to make it clear what remaining steps are needed to achieve an autonomous artificial cell."

By the middle of 2006, however, not even a lowly complemented protocell had been created by any of the PACE member organizations—which was no surprise to anyone involved, given the complexity of the project and the formal timetable for results that the scientists had established at the outset.

Norman Packard, Mark Bedau, and their ProtoLife crew, meanwhile, were pursuing a suite of parallel research agendas, one of which was to implement an evolutionary approach to the design of vesicles, another of which was to orient vesicle design toward targeted drug-delivery systems.

"The possibility of using evolutionary design of complicated, messy systems, involving lots of amphiphiles and other kinds of things, is now established," said Mark Bedau in 2006. "In this coming year we're planning on adding various nucleic acids like PNA and DNA to the mixture, and working on optimizing the interaction between the replicating chemical systems and the container chemical systems as represented in vesicles."

Translated into layman's terms, this meant that they, too, were still very far from actually creating artificial living cells. Additionally, the company had no commercial products on offer, and had not made any sales. ("There's nothing in the window yet," as Steen Rasmussen put it. "The baker's still busy.") On the other hand, extended research and development horizons, along with years of nil returns on investment, were nothing new in the high-tech start-up business, and nobody at ProtoLife was unduly concerned. Nevertheless, Mark Bedau had returned to Reed College on a half-time basis, but spent the other half of the year at his second home in Venice and his second office in Marghera.

At about this same time, in the summer of 2006, Stephen Wolfram was holding a conference in Washington, D.C. Called the "NKS 2006 Wolfram Science Conference" and attended by some hundred or so researchers from various scientific disciplines, it was held at the Fairmont Washington, a glossy, high-end hotel located at the edge of Georgetown. In his keynote address, Wolfram noted that scientific papers on his book A New Kind of Science were now appearing at the rate of two per day; not all of them were positive, of course, but he was clearly having an impact. He also had a word of advice for those working on creating minimal cells.

"One thing NKS suggests is that at some level it should only take quite simple rules to make much of the complexity we see," Wolfram said. "And certainly when people start talking about minimal biological organisms, that's a highly relevant thing to realize."

Norman Packard had known Wolfram as far back as 1986, when both of them were research fellows at the Institute for Advanced Study at Princeton. Packard was at that point a member of Wolfram's complex systems group at the Institute, and so I asked him whether he thought Wolfram's grand plan for reducing the world's complexity to cellular automata programs had any practical relevance to the progress of everyday, hands-on, experimental research at ProtoLife.

"Well, Wolfram is brilliant, so he might be able to find links between simple rules and things like protein folding and the architecture of biological systems," Packard said. "I personally feel the need to understand particular examples of how rules emerge from real interacting components—chemistry and self-assembly in the laboratory."

In other words, no.

Back in Venice, by far the greatest changes had taken place at ECLT—the European Center for Living Technology—which had vacated their old premises in the Palazzo Giovanelli and had moved into not one, not two, but three separate locations in and around the city, all of them offered and supported by the University Ca' Foscari of Venice. First was the Ca' della Zorza, adjacent to the Palazzo Ca' Foscari (the seat of the university), which housed the ECLT's administrative offices. Next was the Palazzo Cavalli Franchetti, a historic palazzo on the Grand Canal near the Accademia Bridge. Franchetti was home to the Venetian Institute of Sciences, Arts and Letters, and in addition housed a conference center and workshop facilities, common rooms, guest rooms, and a garden, all of which the university placed at the disposal of the ECLT.

Still, those palazzi, grand as they were, were as nothing compared to the third and final manifestation of the ECLT, which was on the island of San Servolo, a small rectangular landmass in the southern lagoon across from the main body of the city. The island was remote to the point of being in a realm of its own, a separate dreamscape in an already other-worldly domain.

A Benedictine monastery had been established on San Servolo in the eighth century. Later, the buildings had been converted into a soldiers' hospital and then into an insane asylum. Finally, after a series of renovations and restorations, the island's buildings were turned over to a set of research institutions. Now the ECLT had access to office space, accommodations, a computation center, cafeteria, and library—all of this in the service of a technology that did not even exist as yet.

THE PRIMARY POINT and purpose of the ECLT was to anticipate and assess the ethical, legal, social, religious, and possible other consequences of converting inanimate matter into artificial living entities and then to announce them publicly so that the outside world would not be blindsided by the creation of new life if and when that watershed event actually took place. All this presupposed, however, that the researchers could determine in some clear-cut and definitive fashion when the dividing line between nonliving and living matter had been crossed. It was not exactly obvious, however, what the requirements were for classifying any given chemical structure as "alive."

Here the researchers faced a dilemma. They did not want to establish criteria so strict that an otherwise plausible chemical system was dismissed from consideration for narrowly technical reasons. On the other hand, they did not want to dumb down the definition of life to the point where an obviously lifeless structure was somehow technically classified as "living." Where was the boundary between the two?

Natalio Krasnogor, a computer scientist at the University of Nottingham and a researcher with CHELLnet, a competing chemical cell project, had an idea. In the related field of artificial intelligence, a theoretical decision procedure existed for determining when a given computer system could "think" or was "intelligent." This was an experiment known as the Turing test, named after Alan Turing, who had proposed the notion in 1950. The core notion of the Turing test was an "imitation game" in which a human interrogator submitted written questions to a computer, which would in turn supply written answers. If, on the basis of its replies alone, the computer was indistinguishable from a human being, then Turing considered the computer to be "intelligent" or "thinking."

If such a neatly defined concept worked for artificial intelligence, why shouldn't something analogous work for artificial cells? Over three days in late July 2006, therefore, the ECLT hosted on the island of San Servolo a workshop meant to devise a Turing test for artificial life-forms. The workshop, billed as "A First International Workgroup Meeting for the Systematic Measure of Success in Artificial Chemical Cell

Research," noted in its official announcement that the various international efforts to create synthetic cells were proceeding "in the face of significant uncertainty about the exact nature of life. The definition of 'life' has invoked innumerable seemingly interminable discussions, ranging from the religious to the philosophical and metaphysical. Still today no definition is universally accepted, and the advisability of even proposing definitions is controversial."

The San Servolo meeting would address these issues. A group of fourteen artificial cell researchers gathered together on the island. They included Krasnogor, Norman Packard, Mark Bedau, and Steen Rasmussen, along with David Deamer of the University of California, Santa Cruz, and Martin Hanczyc of ProtoLife, among others, all of whom now set about the task of devising a scheme for determining when an entity qualified as being artificially alive.

"This was hard," Steen Rasmussen recalled afterward. "At least initially there was some divergence in opinion as to what a Turing test for life is."

Many ideas were put forth, some of which would have eliminated from consideration practically everybody who was working on the problem. For example, some people "were looking for very high-level functionalities that required huge, complicated gene networks, where you had communication protocols for signaling molecules, and so on," Rasmussen said. "All of us who are working on these simple systems, we wouldn't even be able to compete!"

There were some extremely imaginative ideas as well; for example, the very Turingesque suggestion that an artificial cell would be alive if it was placed in an environment of conventional, biological cells, and the biological cells couldn't tell whether the newcomer was an artificial cell or a real cell.

"That was of course a computer scientist who came up with that one," said Rasmussen. In fact it had been none other than Nat Krasnogor himself, who still thought it was a good idea. "It was the foundational idea for the Critical Assessment of Artificial Chemical Cells (CRAACC) and a number of papers that are being written," he said in an email after the workshop.

Later, in the October 2006 issue of *Nature Biotechnology*, Krasnogor and eight coauthors published one of those papers: "The Imitation Game—A Computational Chemical Approach to Recognizing Life." Here it became clear what their rationale was for using a Turing test, rather than a formal definition, to determine when the first genuine artificial living chemical cell had been created. One was that no agreed-upon, all-purpose, formal definition of life as yet existed. Indeed, the artificial cell researchers had no more defined life than Turing (or anyone else) had defined thought, which meant that using a definition to recognize life was not even possible in principle.

Nor, secondly, did the workshop attendees want to propose one, the reason being that any candidate definition might inadvertently be too restrictive. Artificial cells, whenever they arrived on the scene, might conceivably be some rather strange beasts, lacking certain common features of biological life while possessing other characteristics not found in conventional cells. Rather than ruling out such specimens, an operational decision procedure for recognizing them was needed. Thus the need for an "imitation game," they wrote, "where the chell [chemical cell] must imitate a natural cell and where an instance of the latter plays the role of interrogator." Natural cells themselves, not humans, would make the decision whether a given artificial cell was alive, on the basis of criteria to be established later.

Separately, the workshop attendees agreed to compile a list of milestones that would mark stages in the journey across the frontier between nonlife and life. Like the Turing test, these milestones would not amount to a formal definition of life, nor a statement of what the essence of life was, but they would at least constitute signposts on the road to final success.

One of those landmark events, the scientists decided, would be the genetic regulation of metabolism, the point at which an artificial cell's metabolic processes were first being controlled by an informational molecule of some sort. Another would be the division of the first cell into two daughter cells, while still another would be the occurrence of evolution among the cells.

To motivate progress toward attaining these goals, the workshop's sponsor, CHELLnet, was offering prize money in the amount of €7,000 (about $10,000) to whichever person or group first reached those turning points or made the most significant experimental advances toward them.

By as early as the spring of 2006, Steen Rasmussen and his group had almost reached the first milestone.

RASMUSSEN'S OFFICE at the Los Alamos National Laboratory was in one of the older buildings on campus, an anonymous-looking, government-issue structure known as building SM-40. The conference room, where he and his staff of some dozen-odd postdocs and senior researchers meet, is a few doors down the hall, which, with its ranks of exposed plumbing, wiring, and steam piping overhead, somewhat resembles the interior of a submarine. In the conference room itself a long table in the middle is flanked by whiteboards on the walls on both sides, science posters interspersed here and there upon them, and, at the front, a movie screen for PowerPoint presentations. The protocell group meets here every Wednesday morning from eleven until twelve-thirty.

"Some of these meetings can get a little exciting," Rasmussen admits.

Still, the meetings, which are devoted to planning, brainstorming, discussion of results and the like, are not where the real work of creating artificial cells gets done. That takes place in the lab, which is in another building altogether and houses their HPLC, a high-performance liquid chromatograph they use to separate various end products; a fluorescence microscope, with which they can see large particles; and the all-important spectrophotometer, used to measure particle sizes. It was in these somewhat retro surroundings that Steen's researchers took some of the first baby steps toward building an operational living protocell.

With an experiment designed by Jim Boncella, one of the project's senior chemists, and the hands-on lab work done by Mike DeClue and other postdocs, Rasmussen and his group managed to get an informational molecule (a "protogene") to control and regulate the metabolic production of container molecules. In other words, they persuaded an artificial gene to direct a chemical reaction that actually built something: a molecule of the kind of fatty acid that would make up the protocell container.

"This wasn't asking the metabolism to do some complicated thing," Rasmussen acknowledges. "It's not like eating peanut butter and then creating skin cells, it's a very simple thing we were asking it to do."

The principle behind it was the so-called photofragmentation reaction, a process in which a molecule captures light energy and, as a result, emits an energy-rich electron.

"This energy-rich electron can then do the chemistry for you: it can digest this resource molecule—that's the idea."

The electron "digests" the molecule in the sense that it breaks it apart, and thereby gives rise to another, separate molecule, which then becomes a component of a larger structure, the container molecule.

This being protocell chemistry (as opposed to conventional biochemistry), some of the molecules involved were rather exotic specimens. Still, what was happening down there among the atoms was in fact a protocellular analogue to the genetic control of metabolism in biology: a gene molecule (8-oxoguanine) mediated a metabolic reaction (the photofragmentation reaction) that in turn produced a body part (the detached portion of the resource molecule that had been split apart by the electron).

Abstractly stated, this all sounded well and good, but there was a catch. The catch was that these reactions were not occurring in whole protocells, baby protocells, or anything remotely approaching them. They were occurring in bulk, in solution, as in a quantity of liquids sloshing around in a test tube (actually a small vial). As such, the experimental medium was the synthetic-biology equivalent of a cell-free biological system, the type of fluidic environment that Marshall Nirenberg had used to decipher the genetic code and that the Rockefeller University researchers Noireaux and Libchaber had put inside their egg-white fatty-acid vesicle bioreactor.

An outsider might view this as halting, incremental progress. Slow as it appeared, however, Rasmussen and his crew were proceeding according to the schedule they had established at the outset.

"We are right on target," he says. "We haven't created life yet but we have demonstrated, we have proven, that it is experimentally feasible to realize this design that we dreamed up many years ago."

For Rasmussen, the hard part was not coming up with theory, doing experiments, or even getting results. The hard part was getting the funding to continue. His main critics, in fact, were those who reviewed his grant requests. "You send in gazillions of applications which are turned down, because people say, 'This can't be done.'"

But as time went on, more and more scientists were trying to do it.

"IT'S NOW pretty clearly a race," said Mark Bedau in mid-2006. By that time, indeed, the protocell project was only one of several competing efforts bent on constructing artificial living cells. Creating life, or something like it, had become one of those once-impossible, now plausible projects—like decoding the human genome—for which the time, the technology, and the science were primed and poised. A clutch of similar programs were under way not only at the Center for Studies in Physics and Biology at Rockefeller University and at CHELLnet (CHELL was an acronym for "chemical cell"), based at the University of Nottingham in the United Kingdom, but also several others, such as the replicating-vesicle project run by Jack Szostak of Harvard Medical School; the Constructive Biology Project at the University of Tokyo; and the supramolecular chemistry project run by Pier Luigi Luisi at ETH Zurich. All told, there were about a dozen international research programs bent on effecting some sort of chemical rapprochement between inert matter and living things.

As an index of how very mainstream the idea had become, there was even a "Cell-Like Entity Project" being run by the United States Air Force. The basic idea, according to the scant information released by the military, was to build an "artificial cell that functions as a human cell would function." Located in Dayton, Ohio, at the Center of Excellence for Cellular Dynamics and Engineering, a research lab run jointly by the Air Force and Wright State University, the program had as its overall goal "to develop revolutionary cell component-based systems for numerous potential military applications, such as monitoring of human health status and detection, identification or neutralization of biological and chemical agents."

How these "cell-like entities" were to do all this was not specified, but what was amply clear from the range and variety of all these efforts was that current science and technology had managed to confront, if not yet to destroy, the distinction between the natural and the artificial within a biological context.

What will it mean, however, if one or more of the scientists working on these projects actually succeeds in creating an artificial living organism?

"There are so many implications of this, ranging from sheer technology all the way down to where we place ourselves in the universe," Mark Bedau said in Venice. "Think about the impact that Darwin had on people's mind-set, where they thought they were in the universe. It just fundamentally shook their view, caused them to re-orient their view of how they fit in the world. And I think making life from scratch will do a similar kind of thing. There will be all sorts of problems and distractions associated with that, people whose cherished beliefs are challenged by this."

Without a doubt there would be some such people, but an alternative response would be equally likely among others. In 1971, when he was president of the American Philosophical Association, the philosopher Lewis White Beck of the University of Rochester delivered the presidential address at the sixty-eighth annual meeting of the association in New York City. The title of his talk was "Extraterrestrial Intelligent Life," and it was about the discovery of possible intelligent life-forms elsewhere in the universe. Such a discovery would be in some sense equivalent to the creation of synthetic life-forms by protocell researchers, and so these two comparable events were philosophically on a par. At the end of his talk, Beck provided an assessment of what the likely impact would be, if any, of the discovery of intelligent life-forms on other worlds.

"I have two conjectures," he said. "First, after a few weeks *it will be forgotten,* just as the details of the first moon landing have already been forgotten by most people. We are so well prepared by popular science and science fiction for signals from outer space that success will be just another nine-days' wonder like Orson Welles's 'Invasion from Mars' or the 'Great Moon Hoax' which shocked New York City in 1837.

"My second conjecture is: *it will never be forgotten.* For what is important is not a single discovery, but the beginning of an endless series of discoveries which will change everything in unforeseeable ways."

Beck did not decide between his two conjectures, and with good reason. As Niels Bohr had said: "Prediction is very difficult, especially about the future." Still, what was true at the time of Beck's talk in 1971 was even truer more than thirty years later, in 2007, by which time people had *already* seen "everything change in unforeseeable ways." They had witnessed not only the moon landing, space stations, artificial hearts, organ transplants, the sequencing of the human genome, gene therapy, cell phones, the Internet, and iPods; they had also witnessed Hiroshima, the collapse of the Twin Towers, the Pacific tsunami, Hurricane Katrina, and much else. It seemed as though people had by now pretty much Seen It All. Against such a roster of major events, the creation of an artificial living cell, if and when it happens, might be just another quaint and curious science item on the nightly news. On the other hand, it could cause an even bigger stir than the debate over genetically modified foods, or evolution by natural selection. There would be no way of knowing until it actually happened.

In any event, a race of living chemical cells *need* not affect people's view of themselves in any fundamental way. For while our concept of what life is would obviously be enlarged by the creation of artificial chemical cells, those cells would not change our conception of *human* life, which would remain exactly what it had been beforehand. Human beings would retain the same value, worth, and uniqueness they had enjoyed previously, and human dignity would not be undercut in the least. After all, *we* would have created *the cells*; *they* would not have created *us*.

In that case, are the creators, the scientists involved, "playing God"?

"Yeah, we *are* playing God, and it's a good thing," Mark Bedau said. "We play God all the time, starting with, you know, agriculture. We try to change the world, including forms of life, in ways that are beneficial. And it's important that we do so, because we've been able to prosper and flourish as a result of it."

Scientists have always meddled with nature and improved upon it. They have changed life, and perhaps someday will even create life.

Still, one thing they haven't done, despite that record of success, is to say what life is. So, what is it?

Ten

What Is Life?

IN HIS SYNOPTIC ENTRY on "Life" that appeared in the 1970 edition of the *Encyclopaedia Britannica*, Carl Sagan wrote that "despite the enormous fund of information that [biologists] have provided, it is a remarkable fact that no general agreement exists on what it is that is being studied. There is no generally accepted definition of life."

One by one, Sagan had reviewed several possible approaches to defining that elusive term. A "physiological" definition interpreted life in terms of typical organismal functions such as eating, excreting, metabolizing, breathing, moving, growing, and so on. "But many such properties are either present in machines that nobody is willing to call alive, or absent from organisms that everybody is willing to call alive," he said. Such a definition included too much: "An automobile, for example, can be said to eat, metabolize, excrete, breathe, move, and be responsive to external stimuli." (The fact that automobiles did many of these things only in some derivative, stretched, figurative, or metaphorical sense did not seem to detract from the force of the example in Sagan's view.) On the other hand, the physiological definition excluded certain living entities: "Some bacteria do not breathe at all, but instead live out their days by altering the oxidation state of sulfur."

Why not a metabolic definition, then, one that saw life as a property of a bounded object that exchanged materials with its surroundings? If there was anything that seemed to be characteristic of any and all living organisms, it was that they engaged in metabolic activities. Still, there were exceptions here, too: there were seeds and spores, Sagan said, "that remain, so far as is known, perfectly dormant and without metabolic activity at low temperatures for hundreds, perhaps thousands, of years, but that can revive perfectly well upon being subjected to more clement conditions."[1] And there were embarrassing false inclusions, such as candle flames that exchanged materials with their surroundings. Were *they* alive?

"Flames also have a well-known capacity for growth," Sagan remarked.

And so on down the list of other possible definitions of life, each of which faced a seemingly insuperable difficulty in the form of one or more counterexamples.

In their book *What Is Life?* Lynn Margulis and Dorion Sagan (Carl Sagan's former wife and their son, respectively) said that life "transcended" any attempt to define it: "Life will self-transcend," they said; "any definition slips away." Whatever that meant.

More recently, Stephen Wolfram came to essentially the same conclusion that Carl Sagan had, and by using basically the same reasoning. In A New Kind of Science, Wolfram had entertained several different criteria for a possible definition of life, and found substantial problems with each of them. Self-organization, for example, sounded like a reasonable defining condition, but the problem was that self-organization was a common enough phenomenon even outside of life: it was found in crystals, snowflakes, weather systems, et cetera.

"In the end," said Wolfram, "every single general definition that has been given both includes systems that are not normally considered alive, and excludes ones that are."

Beyond that was the view of ProtoLife's chief operating officer, the philosopher Mark Bedau, who in a 2004 essay entitled "How to Understand the Question 'What Is Life?'" voiced the possibility that "vital phenomena might have no unified explanation and life might not be a basic category of natural phenomena." While those possibilities might not rule out a universal definition of life, they suggested that there might be no single answer to the "What is life?" question.[2]

Striking the most pessimistic note of all was Édouard Machery, a philosopher of science at the University of Pittsburgh, who in a 2006 paper entitled "Why I Stopped Worrying About the Definition of Life . . . and Why You Should as Well" examined a raft of definitions proposed by scientists and philosophers during the late twentieth and early twenty-first centuries, concluding from this survey that "the project of defining life is either impossible or pointless." Life, Machery said, was either a traditional and ill-defined "folk notion" or a precise, scientific theoretical concept. In the former case, the concept was too vague and diffuse an idea for a formal definition of it to be possible, whereas if it was a precise theoretical notion, then there was an even greater problem, which he called *the embarrassment of riches*. There were a multiplicity of scientific disciplines, and each of them offered its own characteristic definition of life. (Further, some of those definitions included as many as eight to ten or more separate defining criteria that had to be met in order for an entity to be considered alive.) When faced with that wretched excess of competing definitions, how did one meaningfully choose among them? According to Machery, you couldn't: "There is no way to decide between heterogeneous definitions of life." The attempt to do so, therefore, was pointless.

Thus the bottom-line view endorsed by many thinkers who had systematically studied the problem was that attempting to define life was a waste of time. After all, with the time, expense, and mental effort lavished on a futile quest to frame a universal and correct definition of life, you could do something that was actually possible and personally useful, like learning to play the piano.

SO WHY NOT START at the other extreme, with death? Dr. Frankenstein, for one, a character who knew all about creating a living being out of dead body parts, had said: "To examine the causes of life, we must first have recourse to death." Unquestionably, if there was anything that appeared obvious about what it meant to be alive, it was possessing the ability to die.

What, then, is death? Unfortunately, even this was not so

clear as it seemed. Common dictionary definitions, for one thing, were hopeless on the subject:

death (deth) *n*.
1. The act of dying; termination of life.
2. The state of being dead.

And so on.

Even professional bioethicists seemed to be stumped by the question. For example, in 2003 Elysa R. Kooperman, writing in the *American Journal of Bioethics*, spoke of "the lack of consensus on a definition of death."

Is death the disappearance of vital signs, such as heartbeat and breathing? That would be true only of those organisms that possessed hearts and lungs; in the context of all earthly life, including plants, fungi, microorganisms, and cells, creatures endowed with such organs were in the distinct minority. Even in humans, vital signs could disappear temporarily, only to be restored later; there have been false declarations of death, premature burials, and other such medical oddities throughout history.

There were furthermore several different categories of death and maybe even *degrees* of death: there was cardiac death, cardiopulmonary death, apparent death, somatic death, brain death, and legal death. And even within the apparently clear-cut category of brain death, or "death by neurological criteria," there were two different types of death: whole-brain death and brain-stem death.

Were there degrees of being alive as well? Certainly there were degrees of *consciousness*: in the world of clinical medi-

cine there were several intermediate states between full conscious awareness and the total and irreversible lack of it.[3]
Over and above states such as sleep, intoxication, drug-
induced stupors, and the like, there were comas, minimally
conscious states (MCS), vegetative states (VS), persistent vegetative states (PVS), and, lastly, brain death.[4] People in these
states were in their own limbo of semiconsciousness or un-
consciousness and might appear to be "as good as dead,"
which some had argued was true of Terri Schiavo, for example, the Florida woman who had lived for fifteen years in a
persistent vegetative state (PVS) before being allowed to die
in 2005 by the withholding of intravenous liquids and conse-
quent dehydration.

But then there was the case of "the other Terry," Terry
Wallis, who existed in a minimally conscious state (MCS) for
nineteen years before spontaneously awakening in 2003, in a
rehabilitation center in Mountain View, Arkansas, and utter-
ing his first word, which was "Mom." A new brain-scan technique called diffusion tensor imaging showed that Wallis's
brain had apparently grown a mass of new axons, nerve fibers
that conducted impulses between neurons. Later, in 2006, a
British woman in a vegetative state (VS) was found, by means
of functional magnetic resonance imaging (fMRI), to have
brain responses to certain words such as "tennis" and "walk-
ing around in your house." However, because it was not
known whether the imaged brain responses meant the same
thing in normally conscious individuals and in VS patients, it
was not correct to regard this patient as "aware," at least not in
the same sense in which normally conscious persons were
aware. That point was emphasized by many of those who had

considered the case. Nicholas D. Schiff, a neurologist at the Weill Cornell Medical College in New York, for example, said: "It raises a lot of questions. At what level is she conscious? Is she really imagining she is playing tennis? Is it possible to communicate with this person? At this point, this doesn't allow us to make any inferences about where this person's consciousness might be." Kenneth W. Goodman, a bioethicist at the University of Miami, said: "We don't really know what parts of your brain lighting up really mean."

While these and similar cases raised questions concerning the quality of life, and whether life in such compromised states was worth living, such questions were philosophical and not scientific in nature, which meant that there was no objective decision procedure for choosing among the possible answers. (According to news reports, Terry Wallis, once he awoke, reported that he was glad to be alive.) In any event, issues concerning degrees of *consciousness* shed no light on the question of what it meant to be *alive*.

THE ATTEMPT TO BUILD a minimal artificial living cell, by contrast, *did* illuminate that question, for it suggested that there was a category of minimal life in a generic or truly universal sense, one that was not limited to any given life-form, or even to natural, earthly, biological life. This circumstance allowed us to generalize across natural life and synthetic life in an attempt to discover common features, shared attributes that might tell us what life, per se, is.

One common feature was that, whether in cells or in human beings, a minimal living organism was characteristically

surrounded by life-support systems in the form of sophisti-
cated machinery and instrumentation that supplied nutrients
and monitored basic life functions. The minimal synthetic
cell, or bioreactor, created by Noireaux and Libchaber at
Rockefeller University, for example, was supported by mechanisms that kept the object in a solution consisting of feed-
stock molecules in the appropriate concentrations, at the
optimum pressure to force those molecules across the cell
membrane and into the vesicle, and so on. A minimal protocell, likewise, one that's just barely ticking over, would be
bathed in the artificial life-support medium inside a microfluidic system that maintained the cell at the correct temperature, furnished the chemical substances needed for metabolism,
and carried away wastes.

For human beings, there were two analogues to the minimal living systems contemplated by protocell researchers: the
first was that of a fetus (which itself was surrounded by a complex life-support system in the form of the uterus and, of
course, the mother), while the second was that of a brain-dead
patient. (A disanalogy between them was that while the fetus
had the potential of becoming an independent entity with the
prospect of a full life in front of it, the same was not true of the
brain-dead patient. This meant that the prospect of aborting a
fetus was a far more ethically charged issue than turning off a
brain-dead patient's life-support system.) A brain-dead patient
was in fact in some sort of a transitional limbo between life
and death. Mehmet Oz, a New York City heart transplant surgeon, said: "Life and death is not a binary system. In between
life and death is a state of near-death, or pseudo-life."[5]

While a brain-dead patient has irreversibly lost consciousness, brain function, and spontaneous respiration, the heart continues to beat, and for this reason such people are sometimes referred to as "beating-heart cadavers." However, if the victim is an organ donor, then a ventilator (breathing apparatus) and other life-support mechanisms will keep the patient "artificially alive" until the organs are "harvested" or "recovered," which is to say, removed from the body. In this state the donor has not only a heartbeat, a breathing rate, a pulse rate, but also blood pressure and a temperature within normal limits, all of which vital signs are monitored by the usual hospital machinery. Metabolic functions continue (at a depressed rate), and organs other than the brain continue functioning, at least for a time.[6]

At one time doctors thought that brain-dead patients would gradually wither away and "die" shortly after the diagnosis of brain death. However, that view is itself now more or less dead. There are documented reports of brain-dead patients being kept alive for 17 days, 68 days, and for longer periods. Brain-dead pregnant females have been kept alive, in one case for 63 days, in another for 107 days, so that their fetuses could reach a state of maturity sufficient to allow for the successful delivery of healthy babies. Meanwhile, the body lies on a hospital bed, just like a "live" patient, produces urine, is capable of bowel elimination, and is surrounded by the entire range of instrumentation and machinery that would normally attend a recovering patient in an intensive care unit.

When, during the organ recovery procedure, the body is entered by scalpel incision, the wound bleeds just as if the patient were a normal operative specimen. To prevent reflex movements during incisions, neuromuscular blocking drugs are administered to the patient. For that matter, so in some cases is a general anesthetic. "Those who recommend general anesthesia," wrote a British anesthetist in 2003, "would presumably be concerned about the possibility of some residual perception of 'painful' stimuli during the process of organ harvesting."

By almost every outward indication, then, the donor's body is, if not very much alive, then at least *minimally* alive. In other words, while the brain is dead and the person is gone, his or her body remains alive in some tenuous, impaired, and compromised sense. (The brain-dead are not "physically dead," as physicians sometimes claim to reassure friends and relatives of the patient. The physically dead are in mortuaries or cemeteries, where their hearts do not beat, they do not bleed when cut, do not produce urine or give birth to live offspring. The brain-dead are *mentally* dead, not physically dead.)

What all this means is that advanced science and technology have brought some decidedly new concepts to the "What is life?" question, notions that were not available to Erwin Schrödinger when he posed it, including the concepts of degrees of life and especially of minimal life. The phenomenon of minimal life, whether in an artificial cell or a brain-dead body, narrows down the question of what life is in a way that was never before possible. It reduces life to the barest essence, to that which is truly inescapable about and inseparable from a living being.

THERE APPEARS TO BE only a single significant attribute that is held in common by both a minimal synthetic cell and a minimally alive human body: metabolism. Metabolism is an inescapable characteristic of all living things, from the minimally alive to the maximally alive. A reasonable answer to the question "What is life?," then, seems to be: *an embodied metabolism*.[7]

So necessary is metabolism to our concept of life that in 1975 when NASA sent the Viking landers to Mars in an effort to find signs of extraterrestrial life on the planet's surface, what the landers were looking for was not signs of reproduction, nor of evolution, but rather of metabolism.

"Each Viking lander carried three instruments designed to detect the metabolic activities of soil microorganisms," said Norman Horowitz, one of the scientists who analyzed the mission results. "First, the gas-exchange experiment was designed to detect changes in the composition of the atmosphere caused by microbial metabolism. Second, the labeled-release experiment was designed to detect decomposition of organic compounds by soil microbes when they were fed with a nutrient. Third, the pyrolytic-release experiment was designed to detect the synthesis of organic matter in Martian soil from gases in the atmosphere by either photosynthetic or nonphotosynthetic processes."

Even Carl Sagan, who in 1970 had some doubts about defining life in terms of metabolism, wrote in 1980 that "if there are microorganisms in the Martian soil, they must take in food and give off waste gases [as animals do]; or they must take in gases from the atmosphere and, perhaps with the aid of sunlight, convert them into useful materials [as plants do]." That is to say, one way or another even extraterrestrial life-forms must metabolize.

Defining life in terms of metabolism does not in any way negate the importance of reproduction and evolution in life or its processes; it merely refrains from including them as defining conditions, and for the sufficient reason that not all living things reproduce (viz., mules and other sterile species; animals and humans without offspring; individual living organs; brain cells), and because it is at least conceptually possible that a given species, once it reaches a certain level of maturity or complexity, will not evolve further. In fact, it was thought by some until recently, when genetic defenses against malaria were found to be different in Mediterranean and African populations, that human evolution had stopped more than fifty thousand years ago.[8]

Defining life in terms of metabolism, however, opens the door to all the objections that have been previously leveled against that criterion. There is, to start with, the dormant-spore objection. How could metabolism be the essence of life if living entities such as entirely normal and healthy plants can spring forth from spores or seeds that, for all practical purposes, have been "dead" for two hundred or even two thousand years?

A possible answer to this is that while they existed in their dormant states, the seeds and spores were themselves "alive" only in a compromised state, analogous to that in which a minimal chemical cell is alive or a brain-dead patient is alive. Dormancy, in other words, would be the plant equivalent of a biological brain-death state. A second answer is that the dormant seeds or spores are in the even more tenuous life state of suspended animation.

"Suspended animation" is not just a science-fiction term:

recently, such states have been artificially induced for medical purposes, at least in mammals. In 2005, Hasan Alam, a trauma surgeon at Massachusetts General Hospital, completed a series of experiments in which he cooled pigs to 10 degrees Celsius (50°F), removed their blood and replaced it with a substitute, and held the motionless animal in that condition for more than two hours, during which time it had no pulse and no electrical activity in the brain. The animal was to all intents and purposes dead. "You would think so," he admitted. "But you can bring it back."

After he reintroduced the blood and warmed the animal, it sprang back to life. Alam claims a 90 percent success rate with the technique, which he hopes to apply to humans who have suffered traumatic injuries.

An even more portentous possibility is that death might in some cases prove to be a temporary rather than a permanent phenomenon. Suppose a protocell is ultimately developed that metabolizes, reproduces itself, and evolves. Suppose further that all of its life functions and molecular activities could be halted for some significant period of time but could then be restarted so that the cell continued to function afterward exactly as it had before. It would be hard to avoid the conclusion that a minimally living entity was temporarily dead and was then brought back to life. Should that ever happen, it would prompt one of the more profound conceptual reorientations ever to be fostered by science. The same condition— temporary death—might, in the future, also apply to people.

There is a second objection to the claim that life is embodied metabolism, the candle-flame objection. Candle flames "metabolize" in the sense that they exchange materials with their environment, give off waste products, produce heat, and so forth. Better still, they even have the capacity for self-reproduction and growth, just as a full-fledged living entity does.

An obvious reply is that a candle flame is not an *embodied* metabolism, and therefore does not in fact provide an exception to the definition. A second reply is that whereas a metabolism builds things (such as new cellular structures), a candle flame mainly consumes and destroys the candle, meaning that the two are not on a par after all.

An answer of a different type was provided back in 1790, by Antoine Lavoisier, the French chemist. Lavoisier's view was that life was in fact just exactly like a candle flame! "Respiration is a slow combustion of carbon and hydrogen, similar in every way to that which takes place in a lamp or lighted candle," he said. "And, in that respect, breathing animals are active combustible bodies that are burning and wasting away . . . It is the very substance of the animal, the blood, which transports the fuel. If the animal did not habitually replace, through nourishing themselves, what they lose through respiration, the lamp would very soon run out of oil and the animal would perish, just as the lamp goes out when it lacks fuel."

Still, despite Lavoisier's imaginative attempt to equate them, a candle flame and a metabolizing cell or animal are essentially too dissimilar to be comparable.

At this point there is only one remaining barrier to defining life as embodied metabolism: the automobile objection. Is it accurate to regard life as embodied metabolism if cars do many of the things that living entities do? But there are things

that cars do not do: they do not maintain themselves by building new component parts as needed, which metabolism does.
They do not repair themselves, which metabolism also does;
humans perform those functions on their behalf. The whole
point and purpose of metabolism is to *make* something: new
structures, new proteins, energy. Metabolism means synthesis, not merely consumption and destruction. A car, for all
its "eating," "breathing," and "excreting" activities, is nevertheless an essentially lifeless moving object, no more self-
sustaining than a lightbulb.

Defining life as embodied metabolism, then, appears to be
at least as adequate as any other definition of life that has been
offered to date. Given the success of such definitions, this one,
too, might have a rather short half-life. Nevertheless, it seems
to be the most defensible theory we have at the present.

THIS, THEN, IS the situation. Scientists know an enormous
amount about the nature and functioning of life. They know
in exquisite molecular detail what the genes are made of, how
they replicate, synthesize proteins, and regulate life's processes.
They understand to a large degree the fundamental processes
of evolution by natural selection.

Still, there are other things that scientists don't know.
They don't know how life arose, where it came from, or
where it's headed. They don't even know whether, when, or
where any of the artificial cell projects will come to fruition
and present us with the world's first synthetic living entity.

Finally there is the fact that while science has brought the
understanding of life and its workings to an unprecedented

peak of detail and precision, it does not and cannot answer the questions centered on life that many people find most urgent and pressing: those having to do with abortion, euthanasia, responsibility to other species or future generations, the moral status of using embryonic stem cells in research or of cloning human beings, and so on.

But there are divisions of labor in many areas of human activity, and so too it is with science, which is concerned with what is, and morality, which is concerned with what ought (or ought not) to be. Science and technology can keep people artificially alive for virtually arbitrary lengths of time; but the decision of when to remove life support is a moral issue, not a scientific one. And the same is true of the other hot-button issues of the day.

Cruel conclusions, perhaps. Then again . . . that's life.

Notes

One • Birth of a Cell

1. "S.r.l." stands for "Società responsibilità limitata," and is the Italian equivalent of a limited liability company.

Two • Schrödinger

1. The coauthors were N. W. Timofeef-Ressovsky and K. G. Zimmer.

Three • Unlocking the Three Secrets of Life

1. In 1994, on the fiftieth anniversary of the paper's publication, the only surviving member of the three-man team, Maclyn McCarty, wrote: "That it was something less than an immediate sensation is in no small part due to its appearance at a time in the course of World War II, some four months before the onset of the Normandy invasion, when most potential readers of the paper were otherwise engaged."

2. Much later, in 2006, a team of scientists announced in *Nature* that they had found a second-level DNA pattern superimposed on the one discovered by Watson and Crick in 1953.

3. In 2006 Matt Ridley, Crick's biographer, stated that it was "Watson-Crick's argument that the structure of nucleic acids explained heredity," not that it explained all of life.

4. Matthaei, who had been trained as a plant physiologist, left Nirenberg's lab in 1962. He ultimately returned to Germany and became a farmer.

Four • The Fiftieth-Anniversary Coronation and Dismissal

1. At the other extreme were those biologists who viewed the DNA molecule as on a par with the Great Second Coming—or even the First. Gunther Stent, for example, once said: "It was like before and after Christ. Things are either before or after the double helix."

Five • ATP and the Meaning of Life

1. In January 2007, a team of University of California, San Diego, bio-engineering researchers reported that they had developed a working computer simulation of all the biochemical metabolic reactions that occurred in human cells. This effort required the consultation of 1,500 books, review papers, and scientific reports in order to construct a database of 3,300 separate metabolic reactions.

2. www.science.smith.edu/departments/Biology/Bio231/krebs.html.

Six • Origins

1. In 1963 Carl Sagan, Ruth Mariner, and Cyril Ponnamperuma produced ATP molecules in a similar experiment also simulating possible primitive earth conditions. Their report on the project stated: "Results indicate that adenosine triphosphate (ATP) and other nucleoside phosphates can be formed in high yield under simulated primitive earth conditions; e.g., by ultraviolet irradiation of dilute aqueous solutions of adenine, ribose, and ethyl metaphosphate." Given their setup, these results were not surprising.

2. "The argument sounds nutty, and is," said Crick's biographer Matt Ridley in 2006. "But then so are all theories about the origin of life."

3. In 2005, in an attempt to address some of these questions, and to explain how all of it could have been accomplished without divine intervention, Harvard University launched an "Origins of Life in the Universe Initiative." The project was expected to last for some years.

Seven • The Spandrels of San Marco

1. In his 1995 book *Darwin's Dangerous Idea*, the philosopher Daniel Dennett proclaimed that natural selection was "the single best idea anyone has ever had."

2. Alfred Russel Wallace conceived the theory simultaneously with Darwin but did not develop or defend it at the length or at the level of detail remotely approaching Darwin's treatment of the subject. Additionally, in the *Origin of Species*, Darwin notes that his ideas had been anticipated by several others, including one Patrick Matthew, who in 1831 "published his work on 'Naval Timber and Arboriculture,' in which he gives precisely the same view on the origin of species . . . as that enlarged in the present volume."

3. In 1995 Darwin's biographer Janet Browne wrote that while on Galápagos, "[Darwin] felt no sudden need to think about the possibility of evolution. He experienced no legendary moment of revelation . . . The irony was that he did not come to understand the meaning of what he saw until long after the ship sailed away from the archipelago."

4. Public opinion polls show that less than half of all Americans believe in evolution by natural selection. In 2006, a survey of eighteen European countries, the United States, and Japan revealed that only the residents of Turkey are less willing to accept evolution than Americans are.

5. In 2006, Princeton University Press published *The Altruism Equation: Seven Scientists Search for the Origins of Goodness*, by Lee Alan Dugatkin, a biologist at the University of Louisville. A problem whose solution seven scientists are in search of cannot be considered as having been resolved.

6. He died on February 3, 2005, at the age of one hundred.

7. Gould had some Marxist leanings.

8. When another researcher repeated the experiment in response to Gould and Lewontin's criticisms and failed to get the same results, the researcher interpreted this too in adaptationist terms!

9. Although he didn't use the term "spandrel," Darwin himself gave an example of one in the *Origin of Species*: "The sutures in the skulls of young mammals have been advanced as a beautiful adaptation for aiding parturition, and no doubt they facilitate, or may be indispensable for this act," he wrote. "But as sutures occur in the skulls

of young birds and reptiles, which have only to escape from a broken egg, we may infer that this structure has arisen from the laws of growth [whatever they might be]."

10. As did Gould's term "exaptation," coined in 1982 (jointly with Gould's colleague Elisabeth Vrba) to cover useful biological features of nonadaptationist origin.

11. The viewpoint that the universe was a giant computer was not new, however: Edward Fredkin had advanced such a claim in the 1950s. In 2006, Seth Lloyd, an MIT professor of mechanical engineering, proposed the same idea in *Programming the Universe: A Quantum Computer Scientist Takes On the Cosmos*. But because of what some saw as a messianic attempt to reform all of science in his own image, Wolfram elicited some rather extreme responses among critics. One review of NKS, posted on the Internet, was entitled "A Rare Blend of Monster Raving Egomania and Utter Batshit Insanity" (www.math.usf.edu/~eclark/ANKOS_reviews.html).

Eight · The Twilight Zone

1. The term "synthetic biology" had been coined in 2000 by the Stanford University chemistry professor Eric T. Kool, who had rather modest goals in view for the new discipline, envisioning it as a way of systematically exploring the chemical properties and local cellular functioning of various altered nucleic acids.

Ten · What Is Life?

1. In 2006, scientists at the Millennium Seed Bank Project of the Royal Botanic Gardens at Kew, England, planted seeds that had been hidden inside a Dutch merchant's notebook, stored in the National Archives, since 1803; several of the seeds germinated normally. In 2005, a group of Israeli scientists claimed to have grown a date palm from a seed that was two thousand years old.

2. In an earlier work, "The Nature of Life," published in 1996, Bedau proposed that life be defined as "supple adaptation," a capacity to produce novel solutions to problems by means of natural selection. However, (1) other authors denied that the capacity to undergo natural selection was a necessary feature of life; and (2) Bedau himself acknowledged that it was logically possible that some species might "never evolve and adapt."

3. The so-called "Glasgow coma scale" supplied a numerical index of degrees of consciousness; the scores could range from 14 (no impairment) to 3 (consistent with brain death).

4. Whole-brain death is the legal criterion of death in the United States. (Not all countries have a criterion of legal death.)

5. In a 1999 piece appearing in *Anesthesiology*, Gail Van Norman, a University of Washington anesthesiologist, quoted a fictional character, Miracle Max, as saying: "There's a big difference between mostly dead and all dead. Now, mostly dead . . . is slightly alive."

6. However, because of damage to the hypothalamus and the pituitary gland, hormonal balance must be provided artificially.

7. "Embodied" because some sort of barrier is needed to separate a living entity from its surroundings. In his book *Beginnings of Cellular Life: Metabolism Recapitulates Biogenesis*, the biologist Harold J. Morowitz said: "To be an entity, distinguished from the environment, requires a barrier to free diffusion. The necessity of thermodynamically isolating a subsystem is an irreducible condition of life . . . It is the closure of an amphiphilic bilayer membrane into a vesicle that represents discrete transition from nonlife to life."

8. In 2006, by analyzing stretches of the human genome, two University of Chicago geneticists reported finding evidence of recent human evolution (within the last ten thousand years).

Bibliography

One • Birth of a Cell

Holmes, Bob. "Alive! The Race to Create Life from Scratch." *New Scientist* 2486 (Feb. 12, 2005): 28.

Rasmussen, Steen, et al. "Bridging Nonliving and Living Matter." Artificial life press copy, Mar. 26, 2003 (manuscript, 104 pp.). Los Alamos National Laboratory, LA-UR-02-7845.

Rasmussen, Steen, L. Chen, B. Stadler, and P. Stadler. "Proto-organism Kinetics: Evolutionary Dynamics of Lipid Aggregates with Genes and Metabolism." *Origins of Life and Evolution of the Biosphere* 34 (2004): 171.

Rasmussen, Steen, et al. "Transitions from Nonliving to Living Matter." *Science* 303 (Feb. 13, 2004): 963.

Shelley, Mary. *Frankenstein*. New York: Oxford University Press, 1969.

Stroh, Michael. "Life Built to Order." *Popular Science*, Feb. 2003.

Web

www.protocell.org

protocells.lanl.gov

www.protolife.net

www.ees.lanl.gov/staff/steen

134.147.93.66/bmcmyp/Data/PACE/Public

bruckner.biomip.rub.de/bmcmyp/Data/ECLT/Public/events.html

Two • Schrödinger

Dronamraju, Krishna R. "Erwin Schrödinger and the Origins of Molecular Biology." *Genetics* 153 (Nov. 1999): 1071.

Kilmister, C. W., ed. *Schrödinger: Centenary Celebration of a Polymath.* Cambridge: Cambridge University Press, 1987.

Moore, Walter. *Schrödinger: Life and Thought.* Cambridge: Cambridge University Press, 1989.

Schrödinger, Erwin. *What Is Life?: The Physical Aspect of the Living Cell.* Cambridge: University Press, 1945.

Three • Unlocking the Three Secrets of Life

Asimov, Isaac. *Asimov's Biographical Encyclopedia of Science and Technology.* Rev. ed. New York: Doubleday, 1982.

Avery, Oswald T., Colin MacLeod, and Maclyn McCarty. "Studies on the Chemical Nature of the Substance Inducing Transformation of Pneumococcal Types." *Journal of Experimental Medicine* 79 (1944): 137.

Burke, James. *The Pinball Effect: How Renaissance Water Gardens Made the Carburetor Possible.* New York: Little, Brown, 1996.

Crick, Francis. "The Double Helix: A Personal View." *Nature* 243 (Apr. 26, 1974): 766.

[Harden, Victoria.] *Breaking the Genetic Code.* Booklet to accompany an exhibit on the research of Dr. Marshall Nirenberg. Bethesda, Md.: DeWitt Stetten, Jr., Museum of Medical Research, National Institutes of Health, 1988.

Lederberg, Joshua. "Honoring Avery, MacLeod, and McCarty: The Team That Transformed Genetics." *Scientist* 8 (Feb. 21, 1994): 11.

McCarty, Maclyn. "A Retrospective Look at the Discovery of the Genetic Role of DNA." *FASEB Journal* 8 (Aug. 1994): 889.

[Nirenberg, Marshall.] Nirenberg Oral History. National Institutes of Health, May 2006 (manuscript, 391 pp.).

Nirenberg, Marshall. "Historical Review: Deciphering the Genetic Code: A Personal Account." *Trends in Biochemical Sciences* 29 (Jan. 2004).

——. Moscow talk, Aug. 1961. Marshall Nirenberg Papers, National Institutes of Health (manuscript, 3 pp.)

——. "The Genetic Code." Nobel lecture, Dec. 12, 1968.

——. "The Dependence of Cell-Free Protein Synthesis in *E. Coli* upon Naturally Occurring or Synthetic Polyribonucleotides." *Proceedings of the National Academy of Sciences* 47 (Oct. 1961): 1588.

Nirenberg, Marshall, and J. Heinrich Matthaei. "Characteristics and Stabilization of DNAase-Sensitive Protein Synthesis in *E. Coli* Extracts." *Proceedings of the National Academy of Sciences* 47 (Oct. 1961): 1580.

Olby, Robert. "Miescher, Johann Friedrich II." *Dictionary of Scientific Biography*. New York: Scribner's, 1981.

Ridley, Matt. *Francis Crick: Discoverer of the Genetic Code*. New York: HarperCollins, 2006.

Wade, Nicholas. "Scientists Say They've Found a Code Beyond Genetics in DNA." *New York Times*, July 25, 2006.

Watson, James D. *The Double Helix: A Personal Account of the Discovery of the Structure of DNA*. New York: New American Library, 1968.

Watson, James D., and F. H. Crick. "A Structure for Deoxyribose Nucleic Acid." *Nature* 171 (Apr. 25, 1953): 737.

Web

profiles.nlm.nih.gov/JJ (Marshall Nirenberg Papers)

Nobelprize.org

Four • The Fiftieth-Anniversary Coronation and Dismissal

Crow, James F. "Erwin Schrödinger and the Hornless Cattle Problem." *Genetics* 130 (Feb. 1992): 237.

Dyson, Freeman. *Infinite in All Directions*. New York: Harper & Row, 1988.

Margulis, Lynn, and Dorion Sagan. *What Is Life?* Berkeley and Los Angeles: University of California Press, 1995.

Murphy, Michael P., and Luke A. J. O'Neill, eds. *What Is Life?: The Next Fifty Years.* Cambridge: Cambridge University Press, 1995.
  • Manfred Eigen: "What Will Endure of 20th Century Biology?"
  • Stephen J. Gould: "'What Is Life?' as a Problem in History."
  • Stuart A. Kauffman: "'What Is Life?': Was Schrödinger Right?"
  • Eric D. Schneider and James J. Kay: "Order from Disorder."

Five • ATP and the Meaning of Life

Atkins, P. W. *Molecules.* New York: W. H. Freeman, 1987.

Boorstin, Daniel J. *The Discoverers: A History of Man's Search to Know His World and Himself.* New York: Random House, 1983.

Dacome, Lucia. "Living with the Chair: Private Excreta, Collective Health and Medical Authority in the Eighteenth Century." *History of Science* 39 (2001): 467.

Goodsell, David S. *The Machinery of Life.* New York: Springer-Verlag, 1993.

Grmek, M. D. "Santorio Santorio." *Dictionary of Scientific Biography.* New York: Scribner's, 1981.

Holmes, Frederic L. "Krebs, Hans Adolf." *Dictionary of Scientific Biography.* New York: Scribner's, 1981.

"Human Metabolism Recreated in Lab." *BBC News,* Jan. 30, 2007.

Krebs, Hans Adolf. "The Citric Acid Cycle." Nobel lecture, Dec. 11, 1953.

Lane, Nick. *Power, Sex, Suicide: Mitochondria and the Meaning of Life.* Oxford: Oxford University Press, 2005.

Moss, Ralph W. *Free Radical: Albert Szent-Gyorgyi and the Battle over Vitamin C.* New York: Paragon House, 1988.

Web

www.science.smith.edu/departments/Biology/Bio231/krebs.html

Six • Origins

Altman, Sidney. "The RNA World." The Nobel Foundation, 2001.

Cech, Thomas R. "Exploring the New RNA World." The Nobel Foundation, 2004.

Crick, Francis. *Life Itself: Its Origin and Nature.* New York: Simon and Schuster, 1981.

Crick, F.H.C., and L. E. Orgel. "Directed Panspermia." *Icarus* 19 (1973): 341.

Darwin, Charles. Letters. Reprinted in Donald Goldsmith, *The Quest for Extraterrestrial Life*. Mill Valley, CA: University Science Books, 1980.

De Duve, Christian: "The Beginnings of Life on Earth." *American Scientist* (September–October 1995).

Dressler, David, and Huntington Potter. *Discovering Enzymes*. New York: W. H. Freeman, 1991.

Dyson, Freeman. *Origins of Life*. 2nd ed. Cambridge: Cambridge University Press, 1999.

Geison, Gerald L. *The Private Science of Louis Pasteur*. Princeton: Princeton University Press, 1995.

Kauffman, Stuart. *At Home in the Universe: The Search for the Laws of Self-Organization and Complexity*. New York: Oxford University Press, 1995.

Miller, Stanley L. "A Production of Amino Acids Under Possible Primitive Earth Conditions." *Science* 117 (May 15, 1953): 528.

———. "Production of Some Organic Compounds Under Possible Primitive Earth Conditions." *Journal of the American Chemical Society* 77 (May 12, 1955): 235.

Ponnamperuma, Cyril, Carl Sagan, and Ruth Mariner. "Ultraviolet Synthesis of Adenosine Triphosphate Under Possible Primitive Earth Conditions." SAO [Smithsonian Astrophysical Observatory] Special Report No. 128 (1963).

Web

nobelprize.org/chemistry/articles/altman

nobelprize.org/chemistry/articles/cech/index.html

Seven • The Spandrels of San Marco

Brockman, John, ed. *The Third Culture: Beyond the Scientific Revolution*. New York: Simon & Schuster, 1995.

Browne, Janet. *Charles Darwin: A Biography*. Vol. 1: *Voyaging*. Princeton, N.J.: Princeton University Press, 1995.

Darwin, Charles. *The Origin of Species*. 6th ed., 1872. New York: Modern Library, 1993.

Gould, Stephen Jay. *The Structure of Evolutionary Theory*. Cambridge, Mass.: Harvard University Press, 2002.

Gould, Stephen Jay, and Richard C. Lewontin. "The Spandrels of San Marco and the Panglossian Paradigm: A Critique of the Adaptationist Programme." *Proceedings of the Royal Society of London* B 205 (1979): 581.

Hecht, Jeff. "Why Doesn't America Believe in Evolution?" *New Scientist*, Aug. 19, 2006.

Larson, Edward J. *Evolution: The Remarkable History of a Scientific Theory*. New York: Modern Library, 2004.

Mayr, Ernst. *What Evolution Is*. New York: Basic Books, 2001.

Morris, Richard. *The Evolutionists: The Struggle for Darwin's Soul*. New York: Henry Holt, 2001.

Williams, George C. *Adaptation and Natural Selection: A Critique of Some Current Evolutionary Thought*. Princeton, N.J.: Princeton University Press, 1966.

Wolfram, Stephen. *A New Kind of Science*. Champaign, Ill.: Wolfram Media, 2002.

Eight • The Twilight Zone

Berg, Paul, and Maxine Singer. *Dealing with Genes: The Language of Heredity*. Mill Valley, Calif.: University Science Books, 1992.

———. *Genes and Genomes*. Mill Valley, Calif.: University Science Books, 1991.

Cello, Geronimo, Aniko V. Paul, and Eckard Wimmer. "Chemical Synthesis of Poliovirus cDNA: Generation of Infectious Virus in the Absence of Natural Template." *Science* 297 (Aug. 9, 2002): 1016.

Dickey, Christopher. "I Love My Glow Bunny." *Wired* 9 (Apr. 2001).

Ferber, Dan. "Microbes Made to Order." *Science* 303 (Jan. 9, 2004): 158.

Gibbs, W. Wayt. "Synthetic Life." *Scientific American* (May 2004): 75.

Kitamura, Naomi, et al. "Primary Structure, Gene Organization and Polypeptide Expression of Poliovirus RNA." *Nature* 291 (June 18, 1981): 547.

Morton, Oliver. "Life, Reinvented." *Wired* 13 (Jan. 2005).

Noireaux, Vincent, and Albert Libchaber. "A Vesicle Bioreactor as a

Step Toward an Artificial Cell Assembly." *Proceedings of the National Academy of Sciences* 101 (Dec. 21, 2004): 17669.

Pollack, Andrew. "Custom-Made Microbes, at Your Service." *New York Times*, Jan. 17, 2006.

Rawls, Rebecca. "'Synthetic Biology' Makes Its Debut." *Chemical and Engineering News* (Apr. 24, 2000): 49.

Smith, Hamilton O., et al. "Generating a Synthetic Genome by Whole Genome Assembly: phiX174 Bacteriophage from Synthetic Oligonucleotides." *Proceedings of the National Academy of Sciences* 100 (Dec. 23, 2003): 15440.

Tucker, Jonathan. "Gene Machines: The Second Wave." *High Technology*, Mar. 1984, 50.

Tucker, Jonathan, and Raymond A. Zilinskas. "The Promise and Perils of Synthetic Biology." *New Atlantis* (Spring 2006).

Weiss, Rick. "Researchers Create Virus in Record Time." *Washington Post*, Nov. 14, 2003, A10.

Web

www.pbs.org/wgbh/nova/sciencenow/3214/01-collins.html

Nine · The Synthetic Cell Turing Test

Beck, Lewis White. "Extraterrestrial Intelligent Life." Reprinted in *Extraterrestrials: Science and Alien Intelligence*, edited by Edward Regis, Jr. Cambridge: Cambridge University Press, 1985.

Cronin, Leroy, Natalio Krasnogor, et al. "The Imitation Game—A Computational Chemical Approach to Recognizing Life." *Nature Biotechnology* 24 (Oct. 2006): 1203.

Web

www.stephenwolfram.com/publications/talks/nks2006
huey.cs.nott.ac.uk/wiki/index.php/Main_Page (CHELLnet)

Ten · What Is Life?

Bedau, M. A. "How to Understand the Question 'What Is Life?'" In M. Bedau, P. Husbands, T. Hutton, S. Kumar, and H. Suzuki, *Workshop and Tutorial Proceedings, Ninth International Conference on the Simulation and Synthesis of Living Systems* (ALife IX). Boston, 2004.

———. "The Nature of Life." In *The Philosophy of Artificial Life,* edited by Margaret Boden. New York: Oxford University Press, 1996.

Black, Richard. "Seeds 200 Years Old Breathe Again." *BBC News* (Sept. 19, 2006).

Elliot, J. M. "Brain Death." *Trauma* 5 (2003): 23.

Horowitz, Norman. "The Search for Life on Mars." *Scientific American* 237 (Nov. 1977): 52. Reprinted in *The Quest for Extraterrestrial Life: A Book of Readings,* compiled by Donald Goldsmith. Mill Valley, Calif.: University Science Books, 1980.

Kooperman, Elysa R. "The Dead Donor Rule and the Concept of Death." *American Journal of Bioethics* 3 (Winter 2003): 1.

Machery, Édouard. "Why I Stopped Worrying About the Definition of Life . . . and Why You Should As Well." 2006 (manuscript, 41 pp.).

Morowitz, Harold J. *Beginnings of Cellular Life: Metabolism Recapitulates Biogenesis.* New Haven, Conn.: Yale University Press, 1992.

Oropello, John M. "Determination of Brain Death: Theme, Variations, and Possible Errors." *Critical Care Medicine* 32 (2004): 1417.

Puswella, Amal, et al. "Declaring Brain Death: The Neurological Criteria." *Journal of Palliative Medicine* 8 (2005): 640.

Roach, Mary. *Stiff: The Curious Lives of Human Cadavers.* New York: W. W. Norton, 2003.

Sagan, Carl. *Cosmos.* New York: Random House, 1980.

———. "Life." *Encyclopaedia Britannica* (1970).

Trivedi, Bijal. "Suspended Animation: Putting Life on Hold." *New Scientist* 2535 (Jan. 21, 2006).

Van Norman, G. A. "A Matter of Life and Death: What Every Anesthesiologist Should Know About the Medical, Legal, and Ethical Aspects of Declaring Brain Death." *Anesthesiology* 91 (1999): 275.

Voss, Henning U., et al. "Possible Axonal Regrowth in Late Recovery from the Minimally Conscious State." *Journal of Clinical Investigation* 116 (2006): 2005.

Web

www.pitt.edu/~machery/papers/Why%20I%20stopped%20worrying.
pdf

www.afrl.af.mil/news/may06/features/dr_frazier.pdf (Air Force cell-like
entity)

Acknowledgments

I owe the original idea and inspiration for this project to Eric Chinski, editor in chief and vice president at Farrar, Straus and Giroux. I am deeply indebted to him for his confidence that a science writer such as myself could follow, however inadequately, in Schrödinger's footsteps. I am also grateful for his thoughtful comments on two earlier drafts of the manuscript.

Thanks are due to John Brockman and Katinka Matson of Brockman, Inc., for their interest, support, and guidance.

I am hugely indebted to the Alfred P. Sloan Foundation for a grant that materially aided the research and writing of this book. Many thanks to Doron Weber, program officer, for his assistance and counsel.

I must thank Ted Greenwald, senior editor, and Chris Anderson, editor in chief, of *Wired* magazine, for sending me to Venice and

Bonn to investigate ProtoLife, PACE, the ECLT, and the inner workings of the protocell project.

For their help in providing sources, documents, contacts, answers to technical questions, moral support, or other assistance, I would like to thank: Robin Henig; Dr. Michael Kröher (*Manager Magazin*, Hamburg); Louise Paquin and Esther Iglich (McDaniel College); Frank Damazo, M.D.; Mark F. Carr, Ph.D. (Loma Linda University); Barney Stern, M.D. (Department of Neurology, University of Maryland); and Victoria A. Harden (NIH).

Special thanks to Marshall Nirenberg (NIH) for providing me with a complete copy of the Nirenberg Oral History and other documents.

A portion of the manuscript dealing with the protocell project was read by Steen Rasmussen. The entire manuscript was read by Harold Morowitz (Krasnow Institute, George Mason University) and by Robert M. Hazen (Geophysical Laboratory, Carnegie Institution of Washington). I am indebted to them all for their corrections and suggestions, but any errors that might remain are the responsibility of the author.

My warmest thanks go to the Four Protocell Musketeers, Norman Packard, Steen Rasmussen, John McCaskill, and Mark Bedau, who received me graciously and answered my questions in a variety of contexts and venues in the United States and Europe. A talk with Steen Rasmussen aboard a motor launch taking us from Palazzo Giovanelli to Venice's Marco Polo Airport was one of the most memorable (if short) interviews in my experience.

Index